Handbook of Industrial Hydrocarbon Processes

Handbook of Industrial Hydrocarbon Processes

Editor

Puneet Dixit

Handbook of Industrial Hydrocarbon Processes
Edited by **Puneet Dixit**

Printed in 2017

ISBN: 978-1-68117-392-4

Library of Congress Control Number: 2015941581

© 2016 by
SCITUS Academics LLC,
616, Corporate Way, Suite 2, 4766,
Valley Cottage, NY 10989

www.scitusacademics.com

This book contains information obtained from highly regarded resources. Copyright for individual articles remains with the authors as indicated. All chapters are distributed under the terms of the Creative Commons Attribution License, which permits unrestricted use, distribution, and reproduction in any medium, provided the original author and source are credited.

Notice

Reasonable efforts have been made to publish reliable data and views articulated in the chapters are those of the individual contributors, and not necessarily those of the editors or publishers. Editors or publishers are not responsible for the accuracy of the information in the published chapters or consequences of their use. The publisher believes no responsibility for any damage or grievance to the persons or property arising out of the use of any materials, instructions, methods or thoughts in the book. The editors and the publisher have attempted to trace the copyright holders of all material reproduced in this publication and apologize to copyright holders if permission has not been obtained. If any copyright holder has not been acknowledged, please write to us so we may rectify.

Contents

Preface vii

Chapter 1 Novel Electrochemical Treatment of Spent Caustic
from the Hydrocarbon Industry Using Ti/BDD 1

Alejandro Medel, Erika Méndez, José L. Hernández-López,
José A. Ramírez, Jesús Cárdenas, Roberto F. Frausto,
Luis A. Godínez, Erika Bustos1, and Yunny Meas

Chapter 2 Extraction of S- and N-compounds from the Mixture of
Hydrocarbons by Ionic Liquids as Selective Solvents 51

Beata Gabrić, Aleksandra Sander,
Marina Cvjetko Bubalo, and Dejan Macut

Chapter 3 Catalyst Deactivation and Engineering Control for Steam
Reforming of Higher Hydrocarbons in a Novel Membrane
Reformer 81

Zhongxiang Chen, Yibin Yan, and
Said S.E.H. Elnashaie

Chapter 4 Alternate Strategies for Conversion of Waste
Plastic to Fuels 117

Neha Patni, Pallav Shah, Shruti Agarwal, and Piyush Singhal

Chapter 5 Systematic Retrofit Design with Response Surface Method
and Process Integration Techniques: A Case Study for the
Retrofit of A Hydrocarbon Fractionation Plant 137

Aurora Hernández Enríqueza, Michael Binnsb,
and Jin-Kuk Kimb

Chapter 6 Production of Hydrocarbon Liquid by Thermal
Pyrolysis of Paper Cup Waste 183

Bijayani Biswal, Sachin Kumar, and R. K. Singh

Chapter 7 Effect of Soil Texture on Remediation of Hydrocarbons-
contaminated Soil at El-Minia District, Upper Egypt 203
Th. Abdel-Moghny, Ramadan S. A. Mohamed,
E. El-Sayed, Shoukry Mohammed Aly, and Moustafa
Gamal Snousy

Citations 241
Index 245

Preface

Industrial Hydrocarbon Processes will present an analysis of the process steps used to produce industrial hydrocarbons from various raw materials. It is offer a thorough analysis of external factors effecting production such as: cost, availability and environmental legislation. Specific processing operations described in the book include: distillation, thermal cracking and coking, catalytic methods, hydroprocesses, thermal and catalytic reforming, isomerization, alkylation processes, polymerization processes, solvent processes, water removal, fractionation and acid gas removal. This book is arranged in an organized, easy-to-read, and understandable manner and presents the process steps that are required to produce chemicals from various raw materials. It will also assist chemists, engineers, and all manufacturing personnel, even specialists, as it is often possible to translate such general procedures from one discipline to another.

Editor

Novel Electrochemical Treatment of Spent Caustic from the Hydrocarbon Industry Using Ti/BDD

Alejandro Medel[1], Erika Méndez[2], José L. Hernández-López[1], José A. Ramírez[1], Jesús Cárdenas[1], Roberto F. Frausto[1], Luis A. Godínez[1], Erika Bustos[1], and Yunny Meas[1]

[1]Centro de Investigación y Desarrollo Tecnológico en Electroquímica, S.C., Parque Tecnológico, Querétaro-Sanfandila, 76703 Pedro Escobedo, QRO, Mexico
[2]Facultad de Ciencias Químicas, Laboratorio de Investigación Electroquímica, Universidad

ABSTRACT

During the crude oil refining process, NaOH solutions are used to remove H_2S, H_2S_{aq} and sulfur compounds from different hydrocarbon streams. The residues obtained are called "spent caustics." These residues can be mixed with those obtained in other processes, adding to its chemical composition naphthenic acids and phenolic compounds, resulting in one of the most dangerous industrial residues. In this study, the use of electrochemical technology (ET), using BDD with Ti as substrate (Ti/BDD), is evaluated in electrolysis of spent caustic mixtures, obtained through individual samples from different refineries. In this way, the Ti/BDD's capability of carrying out the electrochemical destruction of spent caustics in an acidic medium is evaluated having as key process a chemical pretreatment phase. The potential production of •OHs, as the main reactive oxygen species electrogenerated over Ti/BDD surface, was evaluated in HCl and H_2SO_4 through fluorescence spectroscopy, demonstrating the reaction medium's influence on its production. The results show that the hydrocarbon industry spent caustics can be mineralized to CO_2 and water, driving the use of ET and of the Ti/BDD to solve a real problem, whose potential and negative impact on the environment and on human health is and has been the environmental agencies' main focus.

INTRODUCTION

The aqueous residues produced by the hydrocarbon industry, which involve the use of sodium hydroxide (NaOH) to remove hydrogen sulfide (H_2S), sulfhydric acid (H_2S_{aq}), and sulfur compounds, found in different fractions of crude oil, are known as spent caustics. The process of producing said residues begins when an aqueous solution of NaOH is mixed with a fraction of oil. Although the oil itself can contain sulfur, the process is commonly carried out after the oil fraction is distilled, due to its contact with air, allowing the formation of H_2S, which is very corrosive and difficult to remove. If the H_2S is formed, the NaOH reacts with it to form sodium sulfide, which is water soluble. The spent caustics are also obtained from different and specific processes, whereby they are classified as sulfidic spent caustics (removal of H_2S and mercaptans from hydrocarbons), naphthenics (removal of naphthenic acids from kerosene and diesel), and cresylic spent caustics (removal of organic acids, phenols, cresols, and xylenols). In the case of naphthenics and cresylics, the NaOH reacts with the naphthenic acids, leading to the formation of sodium naphthenates and phenolates, respectively. Accordingly and depending on the quantity and type of products processed, a refinery can generate spent caustics with multiple characteristics containing sulfide (S^{2-}, 1–4 wt%), mercaptans (0.1–4 wt%), phenols (0–2,000 mg L^{-1}), total organic carbon (TOC, 6,000–20,000 mg L^{-1}), chemical oxygen demand (COD, 20,000–60,000 mg L^{-1}), biochemical oxygen demand (BOD, 5,000–15,000 mg L^{-1}), and potentially toxic elements (PTEs) such as Cu, Ni, Cd, Pb, and Cr, among others [1], which require unconventional handling and treatment due to their highly toxic nature and unpleasant odor (the mercaptans can be detected in ppb). In the specific case of spent caustics produced from olefins plants, large amounts of S^{2-} (14,000–21,000 mg L^{-1}), pH values of 13.5 to 13.7, and emulsified hydrocarbons have also been reported [2]. With regard to the phenolic content, depending on the refining process, it can reach up to 30,600 mg L^{-1} [3]. The potential danger of the spent caustics can be understood considering the fact that, by not having technology for its treatment in situ, the different types of spent caustics can be combined, resulting in the production of highly toxic mixtures. On the other hand, when the spent caustics are not treated immediately, these are temporarily stored or confined by authorized companies, with their transport and storage being a latent threat due to the possibility of spillage, which has already been recorded in history with a highly negative impact on human health. A concentration of phenol ranging from 5 to 25 mg L^{-1} is as toxic for aquatic life as it is for humans [4,5]. For this reason, the maximum limits allowed of phenol residues found in industrial discharges vary between 0.5 and 1.0 mg L^{-1} [5]. In the case of potable drinking water, the European Union (EU), in its 80/778/EC Directive, assigned a maximum limit allowed of <0.0005 mg L^{-1} for phenol in all of its forms [4], given that the consumption of water containing these compounds can induce cancer or death [6]. At the same time, the elevated danger of this contaminant for aquatic life has marked phenol and some phenolic compounds as priority contaminants, in agreement with the criteria of the Environmental Protection Agency (EPA) of the United States [4, 5]. In a conventional manner, the treatment of spent caustics from refineries has been carried out in three steps: (1) wet air oxidation (WAO), (2) acid neutralization (AN), and (3) biological oxidation (BO). On occasions, step (2) can be followed by steam stripping. After AN, stripping removes residual H_2S and mercaptans. The liquid effluent obtained has high BOD and COD concentrations because the major portion of the organic constituents is unaffected by the stripping process. Although WAO can treat spent caustic to lower than 1,000 mg L^{-1} COD in 60 min at 475 K and 28 bar, the process is very expensive, and due to severe reaction conditions, safety is a major concern [2]. At the same time, though the elimination of contaminants in steps (1) and (2) can lead to the obtainment of wastewater easily treatable by BO, this depends strictly on the level of conversion and removal of the contaminants found. For example, if the sulfur content is high and the concentration of phenols is low, the BO (use of bacteria from the genus Thiobacillus) is possible. Contrarily, the process can be deactivated when the concentration of phenol is high. It has been reported that effluents containing phenol with a concentration of >3000 mg L^{-1} cannot be treated through BO [4]. Taking this problem into account, different processes have been applied and evaluated for the treatment of effluents from refineries, leaving aside the treatment of the spent caustics. It is important to mention that the physicochemical nature and composition of the spent caustics is not comparable to wastewater produced as part of the extraction processes of crude oil (water produced) or the wastewater produced in other processes. In spent caustics the content of phenolic compounds can reach a value of 30,600 mg L^{-1} [3], while in water produced or any other kind of wastewater this value can oscillate between 20 and 200 mg L^{-1} [7–9]. In the case of wastewaters from refineries, the chemical coagulation using polyaluminium chloride [10], chemical precipitation [11], integrated processes (coagulation-flocculation and flotation) [12], and electrocoagulation (EC) has been reported to remove the high content of the organic material found [13]. At the same time and considering the importance of the phenolic content, as well as the content of commonly found fats and oils [14], the EC has also been evaluated, using NaCl, reaching removal percentages of 91% [15] and 94.5% [16] for synthetic and real samples, respectively. In other studies, the combination of electroflotation (EF) and EC and integrated processes (EC-adsorption-BO) have also been proposed [17, 18]. Although, in spent caustics, the EC has been tested for the removal of sulfides and organic material [1], it is important to highlight that the nature of this process is not destructive, allowing the transference of the pollutants from one phase to another with the subsequent disposal problem. On the other hand and considering the technical difficulties in treating aqueous wastes containing phenol or phenolic compounds through WAO, BO or EC, alternative processes of a destructive nature such as the chemical advanced oxidation processes (AOPs, in the presence or absence of sunlight or assisted light), like the use of hydrogen peroxide (H_2O_2), ozone (O_3), photocatalysis, Fenton process, and the ozonation in alkaline medium, have been amply tested in the treatment of phenols and phenolic compounds. However, the majority of these studies have only been carried out on synthetic samples. In real samples from refineries, applying these processes, only a few studies have been reported [19]. Considering the importance implied in the treatment of real samples, in the last years (2000–2014), different isolated studies have been reported on the treatment of spent caustics using the Fenton process. In this sense, samples containing a COD of 20,160 mg L^{-1} and a total of phenols of 1,800 mg L^{-1} have been successfully treated obtaining removal efficiencies of 90% for COD and 99% for total phenols at a pH of 4 [20]. At the same time, when comparing this process using assisted light (photo-Fenton) in the treatment of sulfidic spent caustics, a greater efficiency was reached with a removal of COD and sulfide up to 97% and 100%, respectively [21]. Applying both processes and despite the excellent results, in practice, the percentage of total destruction of the organic content is strictly dependent on the physicochemical composition of the sample to be treated. An elevated organic charge consumes a large amount of H_2O_2. Also, because of a high concentration of H_2S (up to 20 g L^{-1}), its reaction with ferric ion causes a loss of iron catalyst activity [2]. At the same time, it is necessary to emphasize that the chemical AOPs do not possess the ability to destroy the reaction byproducts. Contrarily, one of the most efficient processes and with the potential to carry out the destruction of any contaminant, including phenol and phenolic compounds, is electrochemical oxidation (EO). Said technology bases its efficiency on the nature of the material used as anode [22]. In this sense, two types of anodic materials are known: (i) active anodes (Au, Ni, stainless steel (SS), Pt, IrO_2, Ti/RuO_2, and analogous combinations) and, (ii) inactive anodes (Ti/PbO_2, Ti/SnO_2-Sb, and analogous combinations), which are materials with a low or high production of •OH radicals (•OHs), respectively. The •OHs can destroy any toxic pollutant to CO_2 and water, due to its high oxidation potential (2.8 V vs. ENH). For this reason the use of inactive anodes is preferred. Here, it is important to mention that Au, Ni, SS, Pt, and analogous combinations are included, due to their similar active or inactive behavior. These materials are not included in the original reference. Although, in the literature, there are numerous studies related to the EO of phenol in synthetic samples [22–53], only a few studies have been reported applying this technology in the treatment of samples from refineries and, as far as we know, there are no reported studies on the treatment of spent caustics. It is also important to consider that both types of aqueous residues represent a highly complex matrix, whose chemical composition can favor or diminish the process efficiency. According to this, using a titanium (Ti) electrode, coated with titanium oxide and ruthenium oxide (RuO_2), efficiencies of phenol degradation in 99.7% and 94.5% and oxidation percentages of COD in 88.9% and 70.1% were obtained for synthetic and real samples (petroleum refinery wastewater), respectively [54]. In other studies using Ti/TiO_2-RuO_2-IrO_2 for phenol degradation in wastewater samples from refineries, removal percentages of 74.75% and 48% for COD and TOC, respectively, were obtained, even when using high quantities of chloride ions [7]. At the same time, the use of Ti/RuO_2-TiO_2-SnO_2 has also been evaluated in the hydrocarbon industry effluents, obtaining low percentages of phenol degradation around 20–47% [55]. Ruling out the use of active electrodes and mixtures of these

[56], a very important option is the use of Ti/SnO$_2$-Sb and Ti/PbO$_2$, which possess a high production of •OHs; however, the application of these materials continues to be restricted due to structural problems and to the possible release of Pb ions [57–59]. As an alternative anodic material, the boron doped diamond (BDD) has been amply evaluated and accepted due to its unusual properties such as a high corrosion resistance, a low adsorption of organic compounds, generation of oxidizing species (O$_3$, H$_2$O$_2$, and •OH) [60], and an elevated overpotential for the reaction of oxygen evolution, ηO$_2$ [61]. The BDD efficiency is based on its high production of •OHs, which are produced through the water oxidation process (1), leading to the complete mineralization of pollutants to CO$_2$ and water (2), with high current efficiencies [62]. This process is known as electrochemical incineration [22]:

$$H_2O \longrightarrow {}^\bullet OH + H^+ + e^- \qquad (1)$$

$$R + M({}^\bullet OH) \longrightarrow M + mCO_2 + nH_2O + H^+ + e^- \qquad (2)$$

It is important to emphasize that different studies on the oxidation of phenol and phenolic compounds in synthetic samples, using BDD, have been presented from 2000 to 2014 [63–86], demonstrating the capability of the use of BDD for the destruction of these pollutants. When comparing the use of Ti/BDD with Ti/SnO$_2$-Sb [87, 88] and PbO$_2$ [89], it was found that the BDD electrode is much better for the destruction of phenol. Although the application of BDD in real samples has been weakly explored, when comparing the BDD with the electro-Fenton process for the treatment of wastewaters from refineries, it has been confirmed that even though the electro-Fenton process can induce a greater phenol degradation in comparison with the BDD electrode, the greater efficiency of degradation of the reaction byproducts (in terms of COD and TOC) occurs with BDD [9, 90]. At the same time, other important studies applying the BDD for the treatment of water produced and typical wastewater from refineries have demonstrated the great efficiency of BDD [91–93]. Said studies and those previously presented not only highlight the effectiveness of the use of the electrochemical technology using BDD but also represent a platform which impulses the use of the BDD, whose scaling potential must be examined in order to advance strategically in solving real problems. According to the above, and based on an extensive review of specialized literature from 1980 to 2014 (use of national and international databases), the object of the present study is to demonstrate the technical feasibility of the use of the Ti/BDD in the electrochemical treatment of spent caustics mixtures, one of the most toxic residues at an industrial level, whose potential and negative impact on the environment and human health is and has been the main focus of environmental agencies.

EXPERIMENTAL DETAILS

Chemicals

Sulfuric acid (H_2SO_4), hydrochloric acid (HCl), NaOH, and phenol (C_6H_5OH) were obtained from J. T. Baker. K_2HPO_4, KH_2PO_4, NH_4OH, potassium ferricyanide ($K_3Fe(CN)_6$), and 4-aminoantipyrine ($C_{11}H_{13}N_3O$) for analysis of phenol in all of its forms were obtained from Aldrich. Coumarin was obtained from Aldrich. The luminescent marine bacteria Vibrio fischeri (Photobacterium phosphoreum) for toxicity analysis was provided by SDI.

Instruments

Analysis of phenol in all of its forms was carried out through visible ultraviolet spectroscopy (UV-Vis), using a Lambda XLS+ spectrophotometer. COD and TOC were evaluated using a Hach model DR/200 reactor/Uv-Vis spectrophotometer DR/2010 and Shimadzu Model TOC-VCSN equipment, respectively. COD was obtained using HACH products. Toxicity analysis was done using a DeltaTox kit, provided by SDI. Electrolysis using ultraviolet light (λ = 254nm) was done by using a Philips mercury lamp with 11 W. Orion Star A215 equipment was used for pH/ORP/conductivity measurements. The electrolysis experiments in galvanostatic mode were performed using Tektronix PWS4323 equipment. The PTEs analyses were performed using a Perkin Elmer Optima 3300 DV model and an Analyst 200/MHS-15 Perkin Elmer. The first equipment was used for the inductively coupled plasma (ICP) analysis and the second for the Hg analyses by hydride generation. The morphology and element analysis of the different electrodes evaluated during the selection of the best anodic material were obtained using a scanning electron microscope (JEOL JMS-6060LV) equipped with an energy dispersive spectrometer (EDS). Crystal structure analysis using X-ray diffraction (XRD) was carried out in a Rigaku Minifles, using Cu K radiation, with a 30 kV operation voltage and 15 mA of current, at a velocity of 2°/min. Cyclic voltammetry (CV) was carried out with an Autolab PGSTAT 30. The •OHs analysis was performed by fluorescence spectroscopy using the equipment HORIBA Jobin MOD Fluorolog 3–22 with double monochromator.

Characterization of the Spent Caustics

Sample Preparation

A total of 10 samples (20 L each) of spent caustic were obtained from different refineries and stored at room temperature. From these samples, a composite sample was prepared by combining 1 L of each individual sample. The sample obtained (called mixture) is shown in Figure 1(a). This process was carried out using safety equipment (Figure 1(b)) and working under an exhaust hood, which was attached to a gas scrubber containing NaOH (Figure 1(c)).

10   Handbook of Industrial Hydrocarbon Processes

(a)

(b)

(c)

Figure 1: Handling and control of (a) spent caustics mixture, (b) safety equipment, and (c) specialized infrastructure.

Before taking each sample, the container was stirred vigorously and the sample was taken immediately. The mixture obtained was also stirred vigorously and stored under refrigeration (4°C) until its analysis.

Physicochemical Analysis

Before carrying out any analysis, the container of the mixture of spent caustics was stirred vigorously. The PTEs analysis was done by using inductively coupled plasma atomic emission spectroscopy (ICP-AES), with the exception of Hg, which was analyzed by hydride generation atomic absorption spectroscopy (HGAAS). The analysis was done prior to digestion of the samples. The concentration of anions present was done through ion chromatography according to the EPA method 300.1 (EPA, 1997). This analysis was performed using a high resolution liquid chromatograph Dionex ICS-2500 HPLC/IC fitted with a Dionex IonPac AS14A column and coupled to a conductivity detector (ED50A). The mobile phase was Na_2CO_3/$NaHCO_3$ (1 mL min^{-1}). The equipment was calibrated using prepared solutions from the 7-Anion Standard of Dionex and the quality of the results was evaluated from the analysis of the certified standard of Inorganic Ventures IC-FAS-1A. The determination of volatile and semivolatile organic compounds (CIDETEQ and Intertek laboratories) in qualitative form was done through gas chromatography mass spectrometry (GC-MS) in accordance with the EPA 5030/EPA 8260C-2006 and EPA 3510/EPA 8270D-2007 procedures. Other physicochemical analyses were done under standard procedures. It is important to highlight that, during the entire process of preparation and analysis of the samples, a strict security protocol was followed, from the use of a personalized infrastructure and safety equipment such as gas masks, nitrile, and neoprene gloves, to that of a special suit.

Electrochemical Treatment

Selection of the Anode

Morphological, structural, and electrochemical analysis of the different materials evaluated (Ti/IrO_2-Ta_2O_5, Ti/SnO_2-Sb, and Ti/BDD) by SEM-EDS, XRD, and CV were done over the same area (2.185 cm^2). Ti/IrO_2-Ta_2O_5 and Ti/SnO_2-Sb electrodes were synthesized by the thermal deposition method using a special formulation. Polycrystalline boron (((B) = 1300 ppm) doped diamond film (Ti/BDD) of 3 μm thickness, provided by Adamant Technologies, was synthesized by hot filament chemical vapor deposition (HF-CVD). The electroactivity of each material was evaluated by CV using a three-electrode cell (60 mL capacity, with a reaction volume of 50 mL). Ti/IrO_2-Ta_2O_5, Ti/SnO_2-Sb, and Ti/BDD electrodes (2.185 cm^2) were used as anodes, a rod of Ti was used as cathode, and a mercury sulfate electrode (Hg/Hg_2SO_4/K_2SO_4

(SAT), V vs. SHE) was the reference electrode. The temperature was maintained at 298 K and the voltammetric profiles for each electrode were obtained applying a scan rate of 100 mV s^{-1} using 0.5 M H_2SO_4 as the supporting electrolyte. Before each analysis, the system was deoxygenated using N_2 gas. In this analysis the Ti/BDD was previously activated (to eliminate C-sp^2 impurities) developing a special methodology [94]. Ti/IrO$_2$-Ta$_2$O$_5$ and Ti/SnO$_2$-Sb were activated by cycling each electrode (50 cycles, 100 mV s^{-1} in 0.5 M H_2SO_4) to stabilize the surface and eliminate impurities.

Selection of the Reaction Medium

Three aspects were taken into account: (i) the pH effect on the chemical state of phenol, (ii) the pH effect on the electrochemical response of the previously selected electrode, and (iii) the pH effect on the production of •OHs. In the first point, analysis by UV-Vis spectroscopy to different pH values was performed with the goal of identifying possible chemical changes on the phenol structure. In this analysis, a synthetic sample was used. The second point was evaluated by the use of CV analyzing two types of acids, HCl and H_2SO_4 (0.5 M), under the same conditions mentioned above. The last point was evaluated through electrolysis experiments, which were performed applying an anodic potential pulse (2.3 V vs. Hg/Hg$_2$SO$_4$, polarization time of 10 min, use of a Pt mesh as counter electrode, 298 K). This last analysis was performed through the use of fluorescence spectroscopy using coumarin (3 × 10^{-5} M) as a probe compound to give way to the formation of the 7-hydroxycoumarin with a wavelength of maximum excitation and emission of 332 nm and 500 nm, respectively ((λ_{ex} = 332 nm, λ_{em} = 500 nm). During electrolysis, samples of 10 μL were withdrawn every minute and immediately analyzed. The equipment and the cell used for detection of the 7-hydroxycoumarin are shown in Figure 2(a).

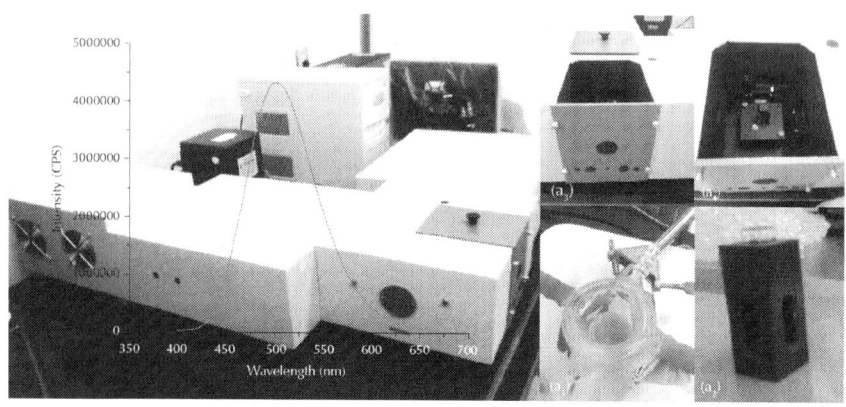

(a)

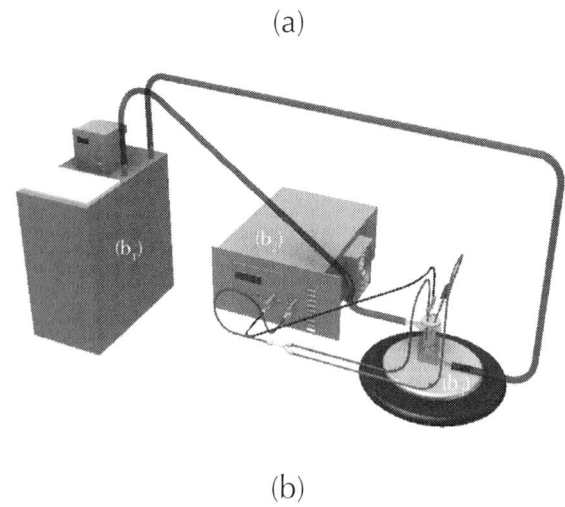

(b)

Figure 2: Experimental system for (a) hydroxyl radicals' analysis by fluorescence spectroscopy, where (a_1)–(a_4) are the steps for analysis of 7-hydroxycoumarin, and (b) electrolysis in galvanostatic mode of spent caustics mixture using Ti/BDD: (b_1) heat exchanger, (b_2) rectifier, and (b_3) electrochemical cell.

Electrolysis of the Spent Caustic

The electrochemical destruction process of the mixture of spent caustics was done in galvanostatic mode, using the experimental system shown in Figure 2(b). In this process, a cell with a single compartment and a capacity of 120 mL (reaction volume of 100 mL, 8 rpm, 298 K and Ti/Pt (3 cm² as counter electrode)) was used. The area of the anode (prior selection) was 3 cm². Electrolysis with UV light (λ = 254 nm) was done under the same conditions indicated above. In each experiment, representative samples were taken at different reaction times (tr), and the degradation of aromatic compounds (phenol and phenolic compounds) and reaction intermediates was done through TOC and COD analysis.

The removal percentages were calculated according to the following formula (3), where C_0 is the initial concentration (mg L^{-1}) and C_f is the final concentration (mg L^{-1}) [90]:

$$\% \text{ removal} = \left[\frac{(C_o - C_f)}{C_o} \right] * 100. \quad (3)$$

The phenol analysis was done using a standard procedure [95], while the toxicity tests were done using a standard method using a lyophilized bacteria, Vibrio Fischeri (Photobacterium phosphoreum), of a luminescent nature, where the reduction of light is proportional to the degree of toxicity.

RESULTS

Characterization of the Spent Caustics

Physicochemical Analysis

Table 1 shows the results obtained in the physicochemical characterization of the sample corresponding to the mixture of spent caustics (mixture of 10 samples). Comparatively, it shows the analysis of a simple sample, corresponding to the sample of greater toxicity (preliminary evaluation of the phenolic content), including the reported values in the literature for spent caustics as well as for wastewater from refineries [1, 3, 7, 9, 13, 16–19, 96–98].

Table 1: Physicochemical analysis of spent caustics

Parameter	Value		Value reference	
	Simple sample	Mixture	Spent caustic	Typical refinery wastewater
Total phenol (mg L^{-1})	7,270.89	11,041.74	30,600 [3]	13 [16], 172.50 [18], 3.17 [17], 192.90 [9], 141 [7], 113 [19], 23 [96]
Oil and grease (mg L^{-1})	239.70	3,399.70	—	85 [97], 1.96 [17], 12.70 [19], 15 [98]
BOD$_5$ (mg L^{-1})	6,930	7,811	—	323 [97], 40.25 [16], 570 [3]
COD (mg O$_2$ L^{-1})	72,065	98,750–102,842.50	72,450 [1], 320,100 [3]	3,150 [97], 100 [16], 4,450 [18], 2,323 [13], 257 [17], 590 [9], 602 [7], 935 [19], 2,746 [98], 1,220 [96]
BOD$_5$/COD	0.09	0.07	—	0.60 [19]
Sulfides (mg L^{-1})	<37.29	<37.29	34,517 [1], 48,500 [3]	19 [19]
Ammonia (mg L^{-1})	—	—	—	13.10 [19]
Cyanides (mg L^{-1})	<0.25	<0.25	—	—
pH	13.02	13.50	12.97 [1], 13 [3]	8 [16], 8.60 [18], 8.05 [13], 7.60 [17], 9.20 [7], 8.10 [19], 7.59 [98], 10 [96]
TOC (mg L^{-1})	—	20,137.50	53,900 [3]	370 [19], 1,500 [96]
Hydrocarbons (mg L^{-1})	—	—	—	11.72 [17], 0.02 [19]
Conductivity (mS cm^{-1})	129.20	208.44	126.70 [1]	6 [18], 13.06 [13], 15.63 [7], 1.73 [7], 0.06 [98]
Alkalinity, CaCO$_3$ (mg L^{-1})	52,266.35	—	15,500 [3]	3,990 [96]
Total dissolved solid (mg L^{-1})	—	169,680	—	5,000 [18], 7,990 [13], 1,333 [7], 4,380 [98]
Settleable solids (mg L^{-1})	—	40	—	—

Total suspended solids (mg L^{-1})	—	3,940	—	22.80 [16], 35 [18], 100 [13], 1,000 [96]
Toxicity (%)	100	100	—	—
F$^-$ (mg L^{-1})	50	135	—	—
Cl$^-$ (mg L^{-1})	86	54,900	37,900 [3]	63 [97], 112 [7], 200,000 [96]
Br$^-$ (mg L^{-1})	12	<0.30	—	—
(mg L^{-1})	11	<0.25	—	—
(mg L^{-1})	381	<0.30	4,600 [3]	—
(mg L^{-1})	608	1,882	20,300 [3]	1,054.50 [13], 212 [7], 1,650 [96]
ORP (mV)	−334.20	—	—	—
Turbidity (NTU)	—	—	—	6.10 [16], 37 [17], 37 [19]

The analysis of the mixture showed a phenolic content of 11,041.74 mg L^{-1}, while the simple sample showed a content of 7,270.89 mg L^{-1}. The difference between both samples clearly demonstrates the environmental problem that commonly appears in the hydrocarbon industry. That is to say, residues of low or high toxicity are mixed with each other, resulting in a residue of greater toxicity. Here, the phenol concentration, organic load, and dissolved solids are not comparable with any other residue, including wastewater from different refining processes [90]. The phenol concentration in the samples analyzed suggest that this residue cannot be treated by BO (a concentration >3,000 mg L^{-1} leads to the complete deactivation of the microorganisms). This is confirmed by the result obtained in the toxicity tests (100% toxic) and through the analysis of the BOD$_5$/COD = 0.09 and 0.07 for the simple sample and the sample corresponding to the mixture, respectively. On the other hand, the elevated alkalinity (52,266.35 mg L^{-1}) and an elevated pH (13–13.5) reflect the reducing nature of these residues, which should be considered in the treatment to be chosen. Many of the contaminants can be fixed at an alkaline pH and be liberated under oxidizing conditions. The reactivity analysis for sulphides and cyanides showed values below the detection limit (<37.29 and <0.25 mg L^{-1}, resp.). Based on the results obtained in this phase of characterization, the samples evaluated were classified as cresylic spent caustics.

Ion Analysis

The analysis of anions, done through ion chromatography, in the case of the simple sample, showed a low content of chlorides (Cl^-) and sulfates (SO_4^{-2}) with values of 86 and 608 mg L^{-1}, respectively, with their concentration being much higher in the mixture, with values of 54;900 and 1,882 mg L^{-1}, respectively. It is necessary to mention that in this case the presence of chlorides is very important. The chlorides, while they can give rise to the presence of compounds of greater toxicity, can also exert a synergic effect during an EO process, giving way to the formation of different oxidizing species [99, 100].

Analysis of Volatile and Semivolatile Organic Compounds

An analysis through GC-MS, corresponding to the mixture of spent caustics, was carried out to determine the different phenolic compounds (reflected in the analysis of phenol in all of its forms, Table 1). Different compounds were considered, of which 43 were for volatile and 62 for semivolatile analysis. The results showed only the presence of 2,4-dimethylphenol, phenol, m-methylphenol, p-methylphenol, and o-methylphenol, all highly toxic compounds. Figure 3 shows the representative chemical structures of each compound identified.

(a)

(b)

(c)

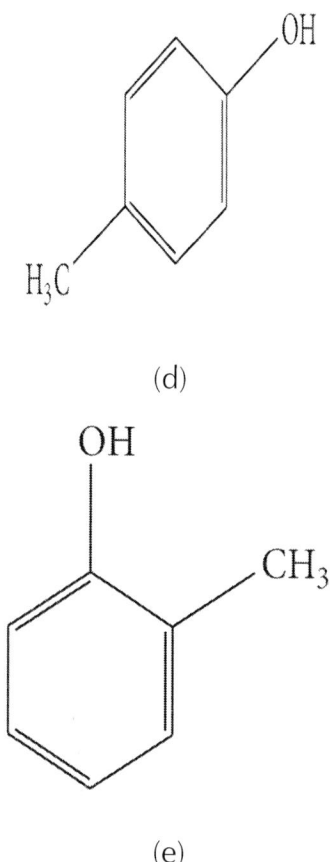

Figure 3: Chemical structure of the main phenolic compounds identified by GC-MS in the spent caustics mixture: (a) 2,4-dimethylphenol, (b) phenol, (c) m-methylphenol, (d) p-methylphenol, and (e) o-methylphenol.

At the same time, a PTEs analysis through ICP was carried out, taking into account that the physicochemical composition of the mixture of spent caustics includes PTEs and possible catalyst traces used in the different processes of hydrocarbon refining. The selection of each PTE was based on an extensive bibliography review. Table 2 shows the main PTEs identified in the mixture sample. In this analysis, a high amount of Na^+ and Fe with values of 19,148 mg L^{-1} and 1,323 mg L^{-1}, respectively, was identified. The presence of Na^+ confirms the caustic nature of the sample, while the presence of Fe, though it is not an especially toxic element, is commonly controlled by the effect of turbidity caused by the precipitation of oxides and hydroxides. The analysis of Fe was done taking into account its catalytic activity on H_2O_2, to give way for the •OH through the Fenton reaction [90]. Due to this, the presence of Fe could be used to accelerate the mineralization process and decrease the residence times during the EO process.

Table 2: Analysis of PTEs in spent caustics mixture

		MLA (mg L^{-1}) in rivers					
PTE (mg L^{-1})		Agricultural irrigation (A)		Public-urban (B)		Protection of aquatic life (C)	
		p.d	p.m	p.d	p.m	p.d	p.m
Al	15.46	—	—	—	—	—	—
As	<0.098	0.2	0.4	0.1	0.2	0.1	0.2
Cd	<0.114	0.2	0.4	0.1	0.2	0.1	0.2
Co	2.42	—	—	—	—	—	—
Cu*	6.89	4.0	6.0	4.0	6.0	4.0	6.0
Cr*	1.95	1.0	1.5	0.5	1.0	0.5	1.0
Fe	1,323	—	—	—	—	—	—
Mn	7.23	—	—	—	—	—	—
Ca	148.87	—	—	—	—	—	—
Mg	34	—	—	—	—	—	—
Na	19,148	—	—	—	—	—	—
Hg*	0.07	0.01	0.02	0.005	0.01	0.005	0.01
Ni*	6.75	2	4	2	4	2	4
Pb*	1.23	0.5	1	0.2	0.4	0.2	0.4
Zn	4.68	10	20	10	10	20	10
V	0.24	—	—	—	—	—	—

On the other hand, PTEs like As and Cd remained under the detection limit (<0.098 and <0.114 mg L^{-1}, resp.), while Cu, Cr, Hg, Ni, and Pb exceeded the maximum limits allowed under the Mexican Regulation [101]. It is important to highlight the elevated toxicity of said elements and, specially, that of the Hg [102, 103] and Pb, which have the ability to enter into the food chains and to concentrate in organisms (magnification process). Considering these analyses, the elevated toxicity of the spent caustics is confirmed, thus justifying the need to direct efforts towards their treatment and/or destruction.

Electrochemical Treatment

Selection of the Anode

Morphology, elemental composition, crystal structure, and electrochemical analysis. The results corresponding to these analyses are shown in Figure 4. The morphology of metal oxide coatings (Ti/IrO$_2$-Ta$_2$O$_5$ and Ti/SnO$_2$-Sb) (Figures 4(a) and 4(b), resp.) showed smaller cracks in comparison with Ti/BDD (Figure 4(c)). The latter showed a compact structure, demonstrating the quality of the coating and the adherence of the diamond to the titanium substrate. In parallel, X-ray dot-mapping was used to analyze the distribution of coating elements.

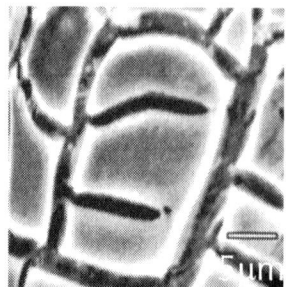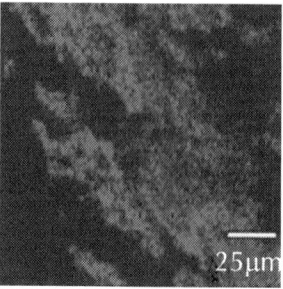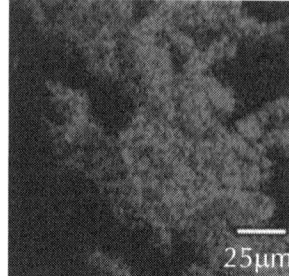

(a)

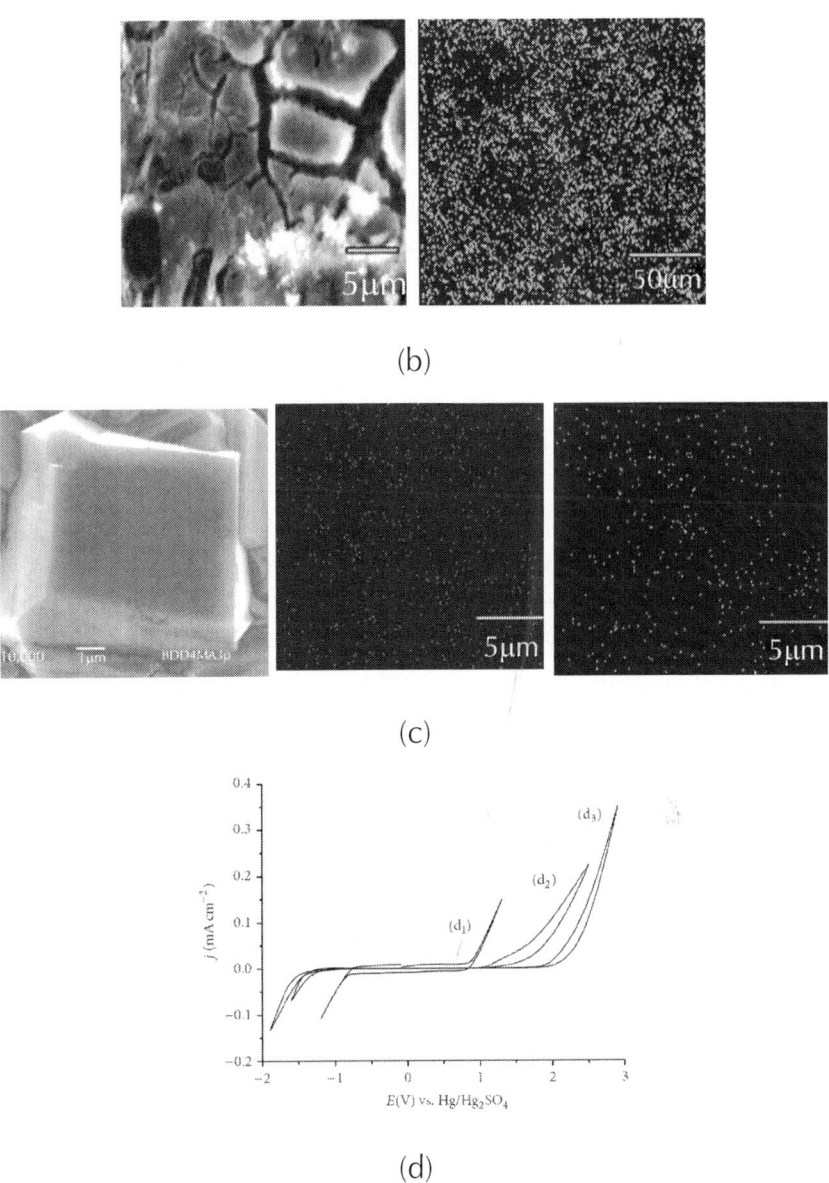

Figure 4: SEM analysis. (a) Ti/IrO$_2$-Ta$_2$O$_5$. (b) Ti/SnO$_2$-Sb. (c) Ti/BDD, applying 15 kV with its respective micrographies and X-ray dot-mapping analysis, identifying: (a$_1$) Ir, (a$_2$) Ta, (b$_1$) Sn, (c$_1$) B, and (c$_2$) C. (d) Cyclic voltammetry analysis for (d$_1$) Ti/IrO$_2$-Ta$_2$O$_5$, (d$_2$) Ti/SnO$_2$-Sb, and (d$_3$) Ti/BDD in 0.5 M H$_2$SO$_4$.

This analysis showed a uniform distribution of Ir (a_1), Ta (a_2), Sn (b_1), B (c_1), and C (c_2). However, the Sb, in the case of Ti/SnO_2-Sb, was not detected. The absence of Sb was probably due to the additive ratio in the coating solution. To verify its presence, an XRD analysis was done demonstrating the presence of SnO_2, Sb_2O_3, and Sb_2O_5. For this specific reason, pretreatment of this electrode was performed before the cyclic voltammetric analysis. At the same time, for both materials (Ti/IrO_2-Ta_2O_5 and Ti/SnO_2-Sb), the wt% of each element was obtained by EDS. In this analysis, the presence of additional elements such as O, Ti, and Si was detected. The results of these analyses (XRD and EDS) have been omitted due to the formulation used. In the case of Ti/BDD, an exhaustive characterization was performed in other studies [94]. The electrochemical characterization of each electrode by CV is shown in Figure 4(d). It shows a comparative analysis of Ti/IrO_2-Ta_2O_5 (d_1), Ti/SnO_2-Sb (d_2), and Ti/BDD (d_3), in 0.5 M H_2SO_4. This graph shows that Ti/BDD has the highest potential window in comparison to Ti/IrO_2-Ta_2O_5 and Ti/SnO_2-Sb. An elevated ηO_2 is important taking into account that the oxidation reaction of the organic compounds during the EO occurs in parallel to the evolution of O_2. Using a material with a high ηO_2, the oxidation reaction is favored over the evolution of O_2, resulting in high current efficiencies. Based on this, it was decided that Ti/BDD was the best material to use in the electrochemical treatment of spent caustic.

Selection of the Reaction Medium

To carry out the electrochemical destruction of the spent caustics using Ti/BDD, the selection of the reaction medium was done taking into account the following approaches: (i) the pH effect on the chemical state of phenol, (ii) the pH effect on the Ti/BDD's electrochemical response, and (iii) the pH effect on •OHs production. According to the first approach, in an extremely alkaline environment (pH 13–13.5, Table 1), the phenol is chemically transformed into sodium phenolates [90]. Although the electronic state of phenol in an alkaline medium has been shown to be favorable for an EO process using electrodes with a low ηO_2 [28], this does not occur using BDD, where the most degradation efficiency is obtained using an acidic medium [90]. Frequently, in literature, different types of acids have been reported for the EO of phenol, such as $HClO_4$ [63,104] and HNO_3 [105]; however, for a possible industrial application, the costs of said acids must be considered. For this reason and considering that during the ion analysis a high quantity of chlorides (54,900 mg L^{-1}) and sulfates (1,882 mg L^{-1}) was identified as part of the chemical composition of the spent caustics mixture, HCl and H_2SO_4 were selected as possible acids to carry out the pH adjustment (13, 13.5 to 1). Initially, the acidification process was evaluated using a synthetic sample in order to rule out any chemical change that occurred on the phenol molecule during the acidification process and to avoid an overestimation of the subsequent results. In said analysis, a solution of 0.05 M NaOH containing phenol 30 mg L^{-1} (TOC) was adjusted to different pH values in an interval of 12–1, using 0.5 M HCl and H_2SO_4. The changes that occurred during the acidification process were evaluated through UV-Vis spectroscopy. Figure 5 shows the different UV-Vis absorption spectrums obtained during the adjustment of the pH using H_2SO_4 (Figure 5(a)) and HCl (Figure 5(b)). It is clearly observed that the chemical conversion of sodium phenolates to phenol (λ = 270 nm) is obtained with both types of acids at a pH of 9, reaching the maximum conversion efficiency at a pH of 7.

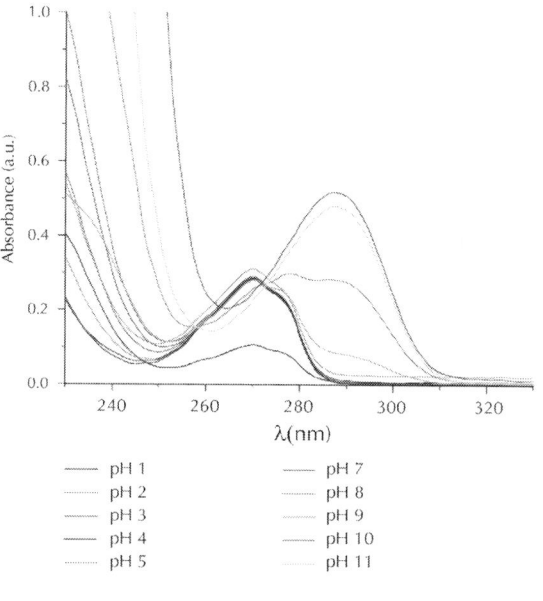

(a)

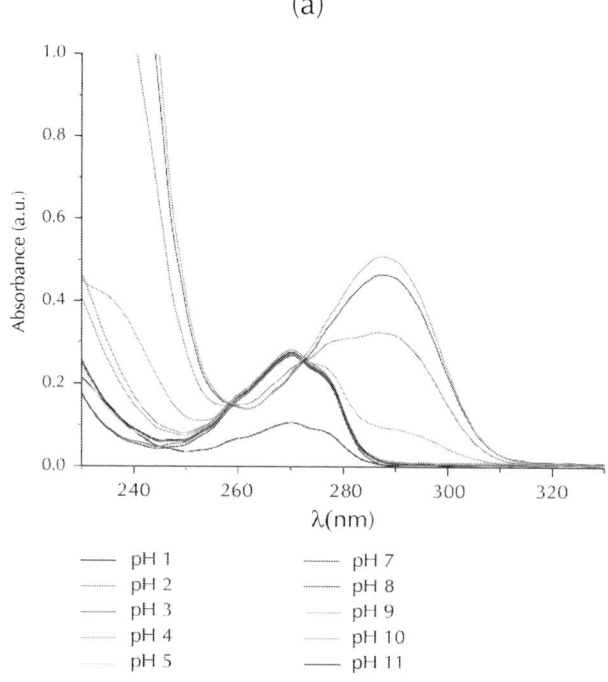

(b)

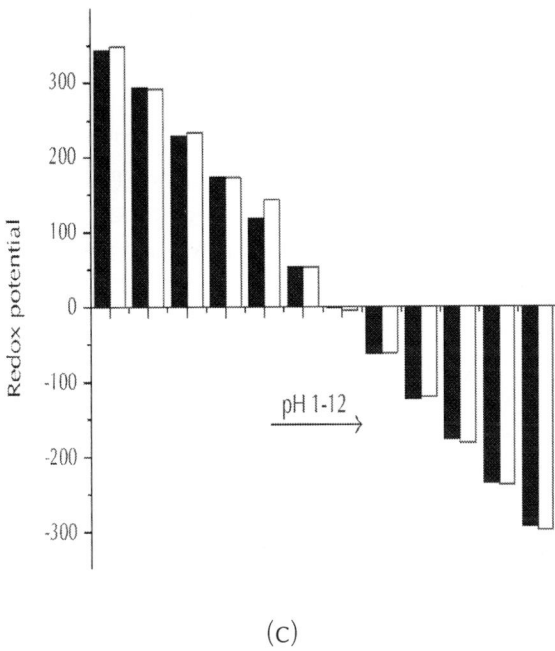

(c)

Figure 5: UV-Vis analysis of the pH effect on the chemical state of phenol, using 30 mg L^{-1} of TOC in (a) 0.05 M NaOH/H$_2$SO$_4$ (pH 12–1), (b) 0.05 M NaOH/HCl (pH 12–1), and (c) ORP analysis under the same conditions to 298 K.

There were no additional and significant chemical changes observed. In parallel to this measurement, an analysis of the redox potential (ORP) was carried out at 298 K. The result of this analysis is shown in Figure 5(c). The values obtained clearly show that an acidic pH favors highly oxidizing conditions, which is of high importance considering that, albeit, an acidic environment does not alter the chemical structure of phenol, the contaminants associated with the spent caustics can be liberated, whereby, a specialized infrastructure and safety equipment are necessary. Using H$_2$SO$_4$, a slight increase in the ORP was observed in comparison with HCl, which is logical considering a greater number of protons. To evaluate the second approach (effect of the pH on the electrochemical response of the Ti/BDD), different voltammograms were obtained through the adjustment of 0.05 M NaOH with 0.5 H$_2$SO$_4$ and HCl in a pH range of 12–1 (data not shown). It was observed that when the pH values descended to acidic, the ηO_2 increased considerably [106]. When comparing the window of potential of H_2SO_4 with that of HCl (pH 1) (Figure 6(a)), a difference in overpotential of 0.5 V was obtained. Considering these results and according to the third approach (pH effect on •OHs production), an analysis of the generation of the •OHs was carried out in both reaction mediums (pH 1). Figure 6(b) shows the influence of the reaction medium on the production of •OHs. It is clearly observed that the production of •OHs is greater in NaOH/H_2SO_4 (pH 1).

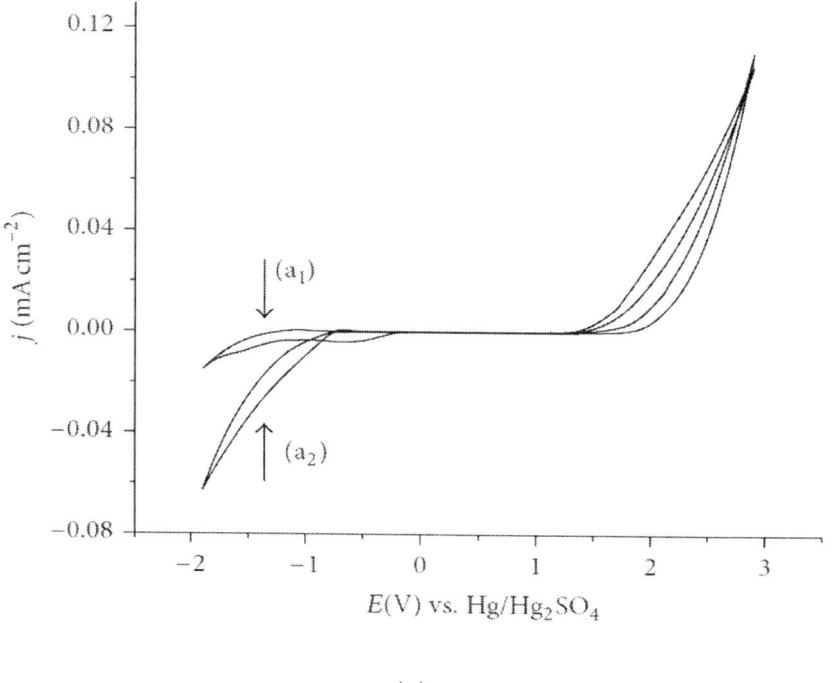

(a)

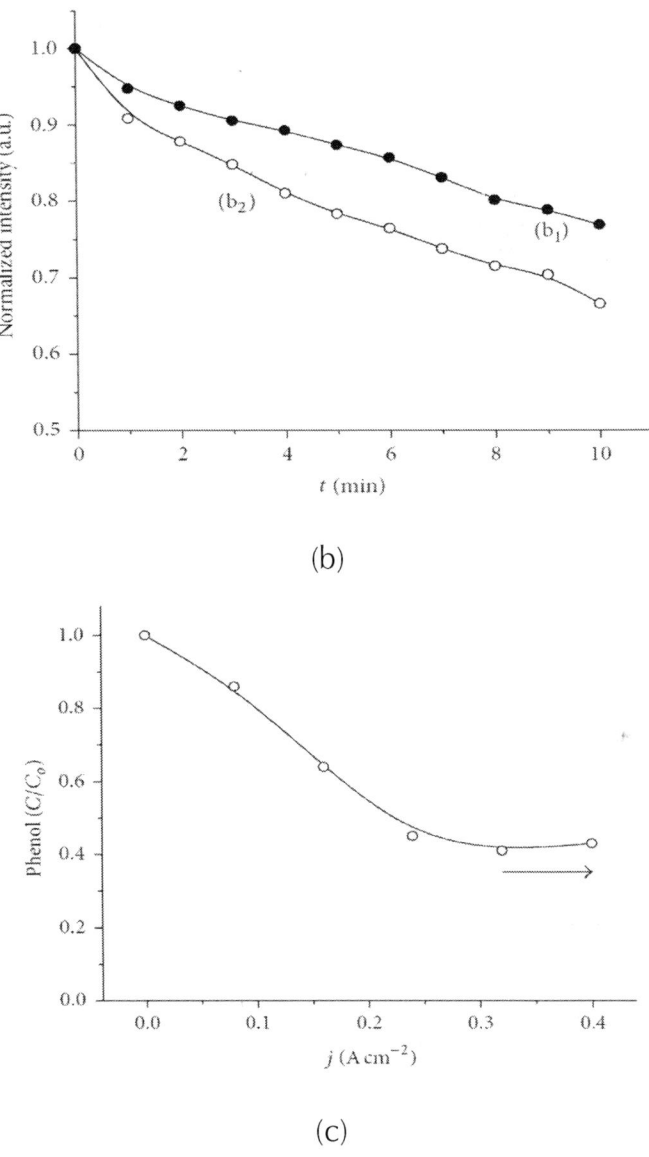

Figure 6: Electrochemical characterization of Ti/BDD by cyclic voltammetric analysis using (a_1) 0.05 M NaOH/HCl (pH 1) and (a_2) 0.05 M NaOH/H_2SO_4 (pH 1). (b) •OHs analysis by fluorescence spectroscopy using (b_1) 0.05 M NaOH/HCl (pH 1) and (b_2) 0.05 M NaOH/H_2SO_4 (pH 1). (c) Selection of the current density using synthetic solutions of phenol (30 mg L^{-1} of TOC) in 0.05 NaOH/H_2SO_4 (pH 1).

This result is of great importance considering that, with the use of a BDD electrode, the formation of any oxidizing species produced in parallel to the oxidation of organic compounds is strictly dependent on the formation of the •OHs. Based on this result, H_2SO_4 was selected to carry out the acidification process.

Electrolysis Using Ti/BDD

Before performing the degradation experiments, a preliminary analysis using a synthetic solution with a phenol concentration of 30 mg L^{-1} of TOC was carried out in NaOH/H$_2$SO$_4$ (pH 1) with the goal of identifying the current to be applied. The currents evaluated were 0.24, 0.48, 0.72, 0.96, 1.20, and 1.44 A, for a 2-hour period, under constant stirring. Ti/BDD (3 cm^2) was used as anode and Ti/Pt (3 cm^2) was used as counter-electrode. It was observed that the middle point of the removal was reached at 0.96 A (j = 0.32 Acm^{-2}), as shown in Figure 6(c). After having identified the current to be applied and before carrying out electrolysis of the spent caustics mixture, a preliminary electrolysis was performed using a simple sample (Table 1). The conditions of electrolysis were $j = 0.32$ Acm^{-2}, pH 1 (adjustment with H$_2$SO$_4$). In this analysis, the effect of ultraviolet light was considered for the purpose of favoring the synergic effect on the production of the •OH. It has been reported that, in the presence of chlorides, the use of ultraviolet light ($\lambda = 254$ nm) can lead to the production of the •OH, according to the reaction (4):

$$HOCl + h\nu \longrightarrow {}^{\bullet}OH + {}^{\bullet}Cl \qquad (4)$$

Figure 7 shows the degradation profiles (TOC and COD) obtained in the different analyses done in the presence and absence of UV light ($\lambda = 254$ nm).

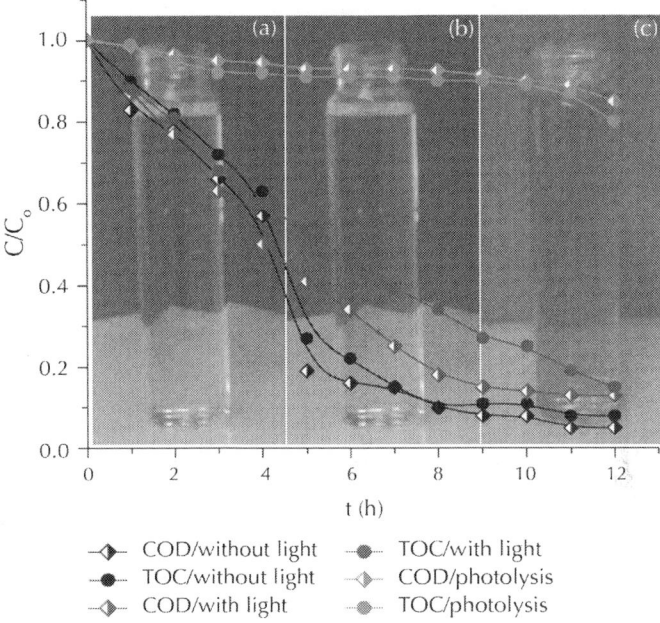

Figure 7: Evaluation of the use of UV light ($\lambda = 254$ nm). Removal of COD and TOC in spent caustic corresponding to the simple sample using Ti/BDD, applying 0.96 A ($j = 0.32$ A cm^{-2}), under constant stirring, pH 1 (adjustment with H_2SO_4) and $t_r = 12$ h. Images of (a) in the absence of UV light, (b) in the presence of UV light ($\lambda = 254$ nm), and (c) photolysis.

The results obtained in electrolysis in the absence of light were more satisfactory than those obtained in its presence ($\lambda = 254$ nm). The results obtained can be attributed to the type of sample analyzed. A real sample is not comparable to a synthetic sample. Here, it is inferred that, due to the high content of organic material, oxidation processes, different from those that occurred in the interface or by the action of the oxidants themselves, are not significant. When analyzing the effect of UV light ($\lambda = 254$ nm), without applying a current (photolysis), no significant change was observed.

Figure 8 shows the analysis of images obtained during the different electrolysis carried out.

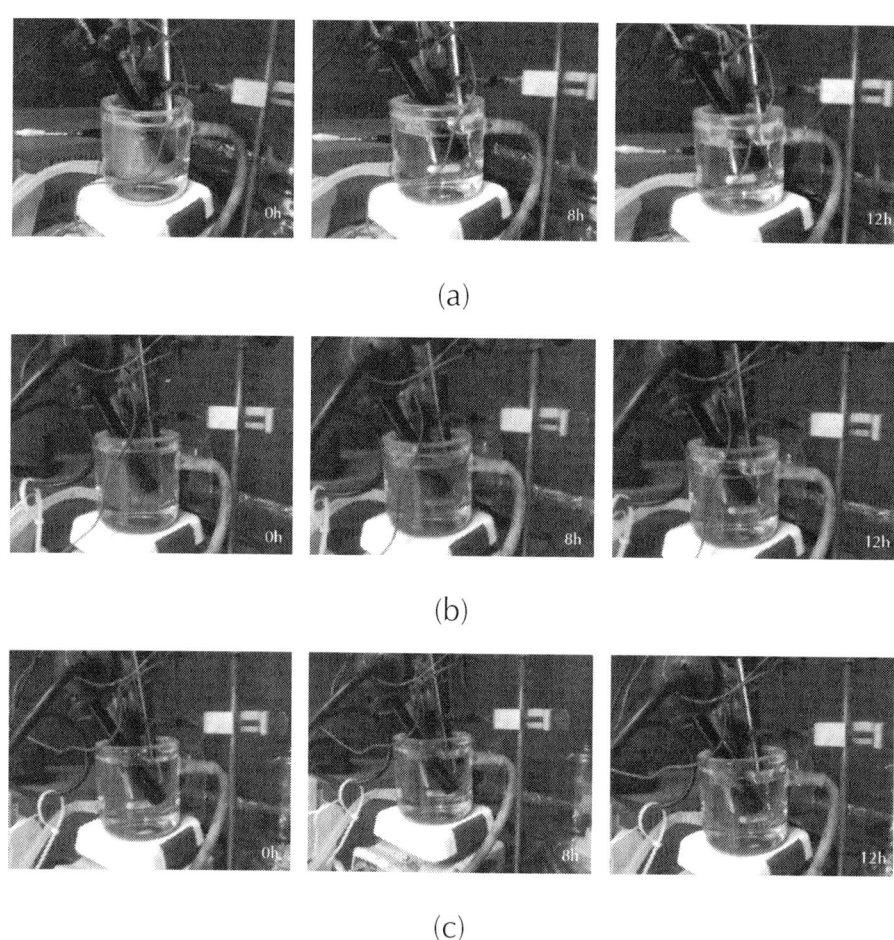

Figure 8: Images of electrolysis of spent caustic corresponding to the simple sample. (a) In the absence of UV light. (b) In the presence of UV light (λ = 254nm). (c) Photolysis, using Ti/BDD, applying 0.96 A (j = 0.32 A cm^{-2}), under constant stirring, pH 1 (adjustment with H_2SO_4) and t_r=12 h.

The initial image (t_r=0h) corresponds to the simple sample submitted to a special pretreatment before electrolysis (acidification to pH 1 using H_2SO_4). A dark brown color in the first stages of the phenol electrolysis is related to the formation of byproducts such as benzoquinone and hydroquinone, known as an active redox couple in equilibrium in an aqueous solution [51]. The color degradation in each experiment was gradual. When comparing each image

(t_r=8h), it is clearly observed that, in electrolysis carried out in the absence of light (Figure 8(a)), the sample turned completely crystalline, indicating a high level of destruction of the organic content. Contrarily, in electrolysis in the presence of light (λ = 254nm) (Figure 8(b)) and in photolysis experiments (Figure 8(c)), the opposite process was observed. In the presence of light (λ = 254nm), the degradation time to obtain a visually crystalline sample was greater (t_r=12h). Based on these results, the use of UV light (λ = 254nm) was discarded. On the other hand and considering the results obtained, the possible passivation of Ti/BDD as an important point was also evaluated. In this analysis, the •OHs production was performed in function of the interfacial potential or current applied. The result (data not shown) indicates that these species only are generated in the zone corresponding to the decomposition of the medium, whereby, if the domains of the potential or current applied are not the correct ones, the electrode can be completely passivated. The image analysis of this test is shown in Figure 9. As can be observed (Figure 9(a)), when working with a current, where the production of the •OHs does not occur, a thick layer of organic compounds is deposited, causing the deactivation of the electrode.

(a)

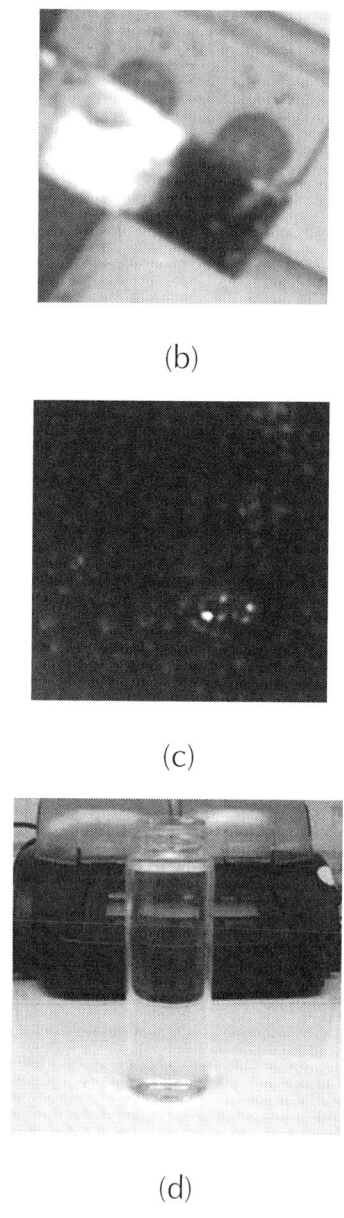

Figure 9: Electrochemical treatment of spent caustics mixture. (a) Applying a current where the •OHs are not generated. (b) Applying a current where the •OHs are generated. (c) Electrolyzed sample (without pretreatment). (d) Sample obtained after the electrochemical treatment with Ti/

BDD applying 0.96 A ($j = 0.32$ A cm^{-2}), under constant stirring, pH 1 (adjustment with H$_2$SO$_4$) and t$_r$=15h.

Contrarily, when the current is applied, inducing a greater interfacial potential by which OHs are electrogenerated, the Ti/BDD electrode can be operated successfully (Figure 9(b)). Finally and according to the previous studies, the electrolysis of the mixture (100 mL) of spent caustics was carried out (pH 1/H$_2$SO$_4$, $j = 0.32$ A cm^{-2}). The results obtained in this analysis are shown in Table 3. The results obtained were similar to those from electrolysis of the simple sample.

Table 3: Electrochemical treatment of spent caustics (mixture) using Ti/BDD

Sample	COD (mg L^{-1})	COD removal (%)	Total removal efficiency, WT-ET (%)	TOC (mg L^{-1})	TOC removal (%)	Total removal efficiency, WT-ET (%)
Without treatment (WT)	98,750	0	—	20,137.5	0	0
Chemical treatment (CT)*	24,533	75.15	—	15,700	22.03	—
Electrochemical treatment (ET)	2,333	90.40	96.63	2,322	85.21	88.46

The analysis of the image of the sample corresponding to the mixture of spent caustics before and after the electrochemical treatment with Ti/BDD (tr=15h) is shown in Figures 9(c) and 9(d), respectively. Complementary analyses of toxicity and phenol in all of its forms showed that the sample electrochemically treated is not toxic, presenting a low phenol content (<3 ppm), with this value being the minimum standard discharge limit for refinery effluents [107].

CONCLUSIONS

A greater production of the •OH was obtained by using H_2SO_4 as a reaction medium. The degradation of spent caustics in an acidic medium (H_2SO_4, pH 1) using Ti/BDD in simple samples proceeded at 100%, while for the mixture, percentages of destruction of 90.40% and 85.21% for COD and TOC, respectively, were obtained. The use of UV light ($\lambda = 254$ nm) did not show an improvement of the process. Further studies are necessary to improve the efficiencies obtained.

Conflict of Interests

The authors declare that there is no conflict of interests regarding the publication of this paper.

ACKNOWLEDGMENTS

The authors would like to thank Mexico's National Council of Science and Technology (CONACyT) for its financial support of this research. The authors also thank Ms. Alejandra Rojo (native speaker) for her review of this paper.

REFERENCES

1. I. Hariz, A. Halleb, N. Adhoum, and L. Monser, "Treatment of petroleum refinery sulfidic spent caustic wastes by electrocoagulation," Separation and Purification Technology, vol. 107, pp. 150–157, 2013.
2. S.-H. Sheu and H.-S. Weng, "Treatment of olefin plant spent caustic by combination of neutralization and fenton reaction," Water Research, vol. 35, no. 8, pp. 2017–2021, 2001.
3. A. Olmos, P. Olguin, C. Fajardo, E. Razo, and O. Monroy, "Physicochemical characterization of spent caustic from the OXIMER process and sour waters from mexican oil refineries,"

Energy & Fuels, vol. 18, no. 2, pp. 302–304, 2004.
4. H. Jiang, Y. Fang, Y. Fu, and Q.-X. Guo, "Studies on the extraction of phenol in wastewater," Journal of Hazardous Materials, vol. 101, no. 2, pp. 179–190, 2003.
5. Y. Han, X. Quan, S. Chen, H. Zhao, C. Cui, and Y. Zhao, "Electrochemically enhanced adsorption of phenol on activated carbon fibers in basic aqueous solution," Journal of Colloid and Interface Science, vol. 299, no. 2, pp. 766–771, 2006.
6. A. Nuhoglu and B. Yalcin, "Modelling of phenol removal in a batch reactor," Process Biochemistry, vol. 40, no. 3-4, pp. 1233–1239, 2005.
7. D. Rajkumar and K. Palanivelu, "Electrochemical treatment of industrial wastewater," Journal of Hazardous Materials, vol. 113, no. 1–3, pp. 123–129, 2004.
8. H. Polat, M. Molva, and M. Polat, "Capacity and mechanism of phenol adsorption on lignite," International Journal of Mineral Processing, vol. 79, no. 4, pp. 264–273, 2006.
9. Y. Yavuz, A. S. Koparal, and Ü. B. Ö ütveren, "Treatment of petroleum refinery wastewater by electrochemical methods," Desalination, vol. 258, no. 1–3, pp. 201–205, 2010.
10. H. Farajnezhad and P. Gharbani, "Coagulation treatment of wastewater in petroleum industry using poly aluminum chloride and ferric chloride," International Journal of Research and Reviews in Applied Sciences, vol. 13, no. 1, pp. 306–310, 2012.
11. L. Alta and H. Büyükgüngör, "Sulfide removal in petroleum refinery wastewater by chemical precipitation," Journal of Hazardous Materials, vol. 153, no. 1-2, pp. 462–469, 2008.
12. C. E. Santo, V. J. P. Vilar, C. M. S. Botelho, A. Bhatnagar, E. Kumar, and R. A. R. Boaventura, "Optimization of coagulation-flocculation and flotation parameters for the treatment of a petroleum refinery effluent from a Portuguese plant," Chemical Engineering Journal, vol. 183, pp. 117–123, 2012.

13. M. H. El-Naas, S. Al-Zuhair, A. Al-Lobaney, and S. Makhlouf, "Assessment of electrocoagulation for the treatment of petroleum refinery wastewater," Journal of Environmental Management, vol. 91, no. 1, pp. 180–185, 2009.
14. C.-L. Yang, "Electrochemical coagulation for oily water demulsification," Separation and Purification Technology, vol. 54, no. 3, pp. 388–395, 2007.
15. M. A. Zazouli, M. Taghavi, and E. Bazrafshan, "Influences of solution chemistry on phenol removal from aqueous environments by electrocoagulation process using aluminium electrodes," Journal of Health Scope, vol. 1, no. 2, pp. 66–70, 2012.
16. O. Abdelwahab, N. K. Amin, and E.-S. Z. El-Ashtoukhy, "Electrochemical removal of phenol from oil refinery wastewater," Journal of Hazardous Materials, vol. 163, no. 2-3, pp. 711–716, 2009.
17. A. Dimoglo, H. Y. Akbulut, F. Cihan, and M. Karpuzcu, "Petrochemical wastewater treatment by means of clean electrochemical technologies," Clean Technologies and Environmental Policy, vol. 6, no. 4, pp. 288–295, 2004.
18. M. H. El-Naas, M. A. Alhaija, and S. Al-Zuhair, "Evaluation of a three-step process for the treatment of petroleum refinery wastewater," Journal of Environmental Chemical Engineering, vol. 2, no. 1, pp. 56–62, 2014.
19. A. Coelho, A. V. Castro, M. Dezotti, and G. L. Sant'Anna Jr., "Treatment of petroleum refinery sourwater by advanced oxidation processes," Journal of Hazardous Materials, vol. 137, no. 1, pp. 178–184, 2006.
20. G. Salas and N. Ale, "Treatment of wastewaters from a pretroleum refinery through advanced oxidation (AOX), using reactive Fenton (H_2O_2/Fe^{2+})," Peruvian Magazine of Chemistry and Chemical Engineering, vol. 11, no. 2, pp. 12–18, 2008.
21. S. Sayid, M. Abu, Z. Noor, S. Noor, and A. Aris, "Fenton and photo-fenton oxidation of sulfidic spent caustic: a comparative study based on statistical analysis," Environmental Engineering

and Management Journal, vol. 13, no. 3, pp. 531–538, 2014.
22. C. Comninellis, "Electrocatalysis in the electrochemical conversion/combustion of organic pollutants for waste water treatment," Electrochimica Acta, vol. 39, no. 11-12, pp. 1857–1862, 1994.
23. V. S. de Sucre and A. P. Watkinson, "Anodic oxidation of phenol for waste water treatment," The Canadian Journal of Chemical Engineering, vol. 59, no. 1, pp. 52–59, 1981.
24. B. Fleszar and J. Poszy ska, "An attempt to define benzene and phenol electrochemical oxidation mechanism," Electrochimica Acta, vol. 30, no. 1, pp. 31–42, 1985.
25. H. Sharifian and D. W. Kirk, "Electrochemical oxidation of phenol," Journal of the Electrochemical Society, vol. 113, no. 5, pp. 921–924, 1986.
26. I. F. McConvey, K. Scott, J. M. Henderson, and A. N. Haines, "Electrochemical reaction with parallel reversible surface adsorption: interpretations of the kinetics of anodic oxidation of aniline and phenol to carbon dioxide," Chemical Engineering and Processing: Process Intensification, vol. 22, no. 4, pp. 231–235, 1987.
27. D.-T. Chin, N. R. K. Vilambi, and C. Y. Cheng, "Oxidation of phenol to benzoquinone in a CSTER with modulated alternating voltage," Journal of Applied Electrochemistry, vol. 19, no. 3, pp. 459–461, 1989.
28. C. Comninellis and C. Pulgarin, "Anodic oxidation of phenol for waste water treatment," Journal of Applied Electrochemistry, vol. 21, no. 8, pp. 703–708, 1991.
29. R. Kötz, S. Stucki, and B. Carcer, "Electrochemical waste water treatment using high overvoltage anodes. Part I: physical and electrochemical properties of SnO_2 anodes," Journal of Applied Electrochemistry, vol. 21, no. 1, pp. 14–20, 1991.
30. S. Stucki, R. Kötz, B. Carcer, and W. Suter, "Electrochemical waste water treatment using high overvoltage anodes. Part II: anode performance and applications," Journal of Applied Electrochemistry, vol. 21, no. 2, pp. 99–104, 1991.

31. O. J. Murphy, G. D. Hitchens, L. Kaba, and C. E. Verostko, "Direct electrochemical oxidation of organics for wastewater treatment," Water Research, vol. 26, no. 4, pp. 443–451, 1992.
32. C. Comninellis and C. Pulgarin, "Electrochemical oxidation of phenol for wastewater treatment using SnO_2, anodes," Journal of Applied Electrochemistry, vol. 23, no. 2, pp. 108–112, 1993.
33. K. T. Kawagoe and D. C. Johnson, "Electrocatalysis of anodic oxygen-transfer reactions. Oxidation of phenol and benzene at bismuth-doped lead dioxide electrodes in acidic solutions," Journal of the Electrochemical Society, vol. 141, no. 12, pp. 3404–3409, 1994.
34. C. Comninellis and A. Nerini, "Anodic oxidation of phenol in the presence of NaCl for wastewater treatment," Journal of Applied Electrochemistry, vol. 25, no. 1, pp. 23–28, 1995.
35. N. B. Tahar and A. Savall, "Mechanistic aspects of phenol electrochemical degradation by oxidation on a Ta/PbO_2 anode," Journal of the Electrochemical Society, vol. 145, no. 10, pp. 3427–3434, 1998.
36. U. Schümann and P. Gründler, "Electrochemical degradation of organic substances at PbO_2 anodes: monitoring by continuous CO_2 measurements," Water Research, vol. 32, no. 9, pp. 2835–2842, 1998.
37. N. B. Tahar and A. Savall, "Electrochemical degradation of phenol in aqueous solution on bismuth doped lead dioxide: a comparison of the activities of various electrode formulations," Journal of Applied Electrochemistry, vol. 29, no. 3, pp. 277–283, 1999.
38. P. Cañizares, J. A. Domínguez, M. A. Rodrigo, J. Villaseñor, and J. Rodríguez, "Effect of the current intensity in the electrochemical oxidation of aqueous phenol wastes at an activated carbon and steel anode," Industrial & Engineering Chemistry Research, vol. 38, no. 10, pp. 3779–3785, 1999.
39. J. Iniesta, J. González-García, E. Expósito, V. Montiel, and

A. Aldaz, "Influence of chloride ion on electrochemical degradation of phenol in alkaline medium using bismuth doped and pure PbO_2 anodes," Water Research, vol. 35, no. 14, pp. 3291–3300, 2001.

40. G. A. Bogdanovskii, T. V. Savel'eva, and T. S. Saburova, "Phenol conversions during electrochemical generation of active chlorine," Russian Journal of Electrochemistry, vol. 37, no. 8, pp. 865–869, 2001.

41. Z. Wu and M. Zhou, "Partial degradation of phenol by advanced electrochemical oxidation process," Environmental Science and Technology, vol. 35, no. 13, pp. 2698–2703, 2001.

42. M. H. Zhou, Z. C. Wu, and D. H. Wang, "A novel electrocatalysis method for organic pollutants degradation," Chinese Chemical Letters, vol. 12, no. 10, pp. 929–932, 2001.

43. R. T. Pelegrini, R. S. Freire, N. Duran, and R. Bertazzoli, "Photoassisted electrochemical degradation of organic pollutants on a DSA type oxide electrode: process test for a phenol synthetic solution and its application for the E1 bleach Kraft mill effluent," Environmental Science and Technology, vol. 35, no. 13, pp. 2849–2853, 2001.

44. M.-H. Zhou, Z.-C. Wu, and X.-D. Xuan, "Anodic-cathodic electrocatalytic degradation of phenol with oxygen sparged in the presence of iron (II)," Chemical Research in Chinese Universities, vol. 18, no. 3, pp. 262–266, 2002.

45. R. L. Pelegrino, R. A. Di Iglia, C. G. Sanches, L. A. Avaca, and R. Bertazzoli, "Comparative study of commercial oxide electrodes performance in electrochemical degradation of organics in aqueous solutions," Journal of the Brazilian Chemical Society, vol. 13, no. 1, pp. 60–65, 2002.

46. Y. J. Feng and X. Y. Li, "Electro-catalytic oxidation of phenol on several metal-oxide electrodes in aqueous solution," Water Research, vol. 37, no. 10, pp. 2399–2407, 2003.

47. P. D. P. Alves, M. Spagnol, G. Tremiliosi-Filho, and A. R.

de Andrade, "Investigation of the influence of the anode composition of DSA-type electrodes on the electrocatalytic oxidation of phenol in neutral medium," Journal of the Brazilian Chemical Society, vol. 15, no. 5, pp. 626–634, 2004.

48. P. Cañizares, J. García-Gómez, J. Lobato, and M. A. Rodrigo, "Modeling of wastewater electro-oxidation processes part I. General description and application to inactive electrodes," Industrial & Engineering Chemistry Research, vol. 43, no. 9, pp. 1915–1922, 2004.

49. P. Cañizares, J. García-Gómez, J. Lobato, and M. A. Rodrigo, "Modeling of wastewater electro-oxidation processes part II. Application to active electrodes," Industrial & Engineering Chemistry Research, vol. 43, no. 9, pp. 1923–1931, 2004.

50. D. Fino, C. C. Jara, G. Saracco, V. Specchia, and P. Spinelli, "Deactivation and regeneration of Pt anodes for the electrooxidation of phenol," Journal of Applied Electrochemistry, vol. 35, no. 4, pp. 405–411, 2005.

51. X.-Y. Li, Y.-H. Cui, Y.-J. Feng, Z.-M. Xie, and J.-D. Gu, "Reaction pathways and mechanisms of the electrochemical degradation of phenol on different electrodes," Water Research, vol. 39, no. 10, pp. 1972–1981, 2005.

52. M. Li, C. Feng, W. Hu, Z. Zhang, and N. Sugiura, "Electrochemical degradation of phenol using electrodes of Ti/RuO_2-Pt and Ti/IrO_2-Pt," Journal of Hazardous Materials, vol. 162, no. 1, pp. 455–462, 2009.

53. E. Chatzisymeon, S. Ferro, I. Karafyllis, D. Mantzavinos, N. Kalogerakis, and A. Katsaounis, "Anodic oxidation of phenol on Ti/IrO_2 electrode: experimental studies," Catalysis Today, vol. 151, no. 1-2, pp. 185–189, 2010.

54. Y. Yavuz and A. S. Koparal, "Electrochemical oxidation of phenol in a parallel plate reactor using ruthenium mixed metal oxide electrode," Journal of Hazardous Materials, vol. 136, no. 2, pp. 296–302, 2006.

55. A. M. Z. Ramalho, C. A. Martínez-Huitle, and D. R. D. Silva,

"Application of electrochemical technology for removing petroleum hydrocarbons from produced water using a DSA-type anode at different flow rates," Fuel, vol. 89, no. 2, pp. 531–534, 2010.

56. M. R. G. Santos, M. O. F. Goulart, J. Tonholo, and C. L. P. S. Zanta, "The application of electrochemical technology to the remediation of oily wastewater," Chemosphere, vol. 64, no. 3, pp. 393–399, 2006.

57. F. Montilla, E. Morallón, and J. L. Vázquez, "Evaluation of the electrocatalytic activity of antimony-doped tin dioxide anodes toward the oxidation of phenol in aqueous solutions," Journal of the Electrochemical Society, vol. 152, no. 10, pp. B421–B427, 2005.

58. Z.-C. Wu, M.-H. Zhou, Z.-W. Huang, and D.-H. Wang, "Electrocatalysis method for wastewater treatment using a novel beta-lead dioxide anode," Journal of Zhejinag University Science, vol. 3, no. 2, pp. 194–198, 2002.

59. M. Zhou, Q. Dai, L. Lei, C. Ma, and D. Wang, "Long life modified lead dioxide anode for organic wastewater treatment: electrochemical characteristics and degradation mechanism," Environmental Science and Technology, vol. 39, no. 1, pp. 363–370, 2005. · ·

60. P.-A. Michaud, M. Panizza, L. Ouattara, T. Diaco, G. Foti, and C. Comninellis, "Electrochemical oxidation of water on synthetic boron-doped diamond thin film anodes," Journal of Applied Electrochemistry, vol. 33, no. 2, pp. 151–154, 2003. ·

61. A. Kraft, M. Stadelmann, and M. Blaschke, "Anodic oxidation with doped diamond electrodes: a new advanced oxidation process," Journal of Hazardous Materials, vol. 103, no. 3, pp. 247–261, 2003.

62. B. Marselli, J. García-Gómez, P.-A. Michaud, M. A. Rodrigo, and C. Comninellis, "Electrogeneration of hydroxyl radicals on boron-doped diamond electrodes," Journal of the Electrochemical Society, vol. 150, no. 3, pp. D79–D83, 2003.

63. J. Iniesta, P. A. Michaud, M. Panizza, G. Cerisola, A. Aldaz, and C. Comninellis, "Electrochemical oxidation of phenol at boron-doped diamond electrode," Electrochimica Acta, vol. 46, no. 23, pp. 3573–3578, 2001.
64. M. Panizza, P. A. Michaud, G. Cerisola, and C. Comninellis, "Electrochemical treatment of wastewaters containing organic pollutants on boron-doped diamond electrodes: prediction of specific energy consumption and required electrode area," Electrochemistry Communications, vol. 3, no. 7, pp. 336–339, 2001.
65. P. L. Hagans, P. M. Natishan, B. R. Stoner, and W. E. O'Grady, "Electrochemical oxidation of phenol using boron-doped diamond electrodes," Journal of the Electrochemical Society, vol. 148, no. 7, pp. E298–E301, 2001.
66. P. Cañizares, M. Díaz, J. A. Domínguez, J. García-Gómez, and M. A. Rodrigo, "Electrochemical oxidation of aqueous phenol wastes on synthetic diamond thin-film electrodes," Industrial & Engineering Chemistry Research, vol. 41, no. 17, pp. 4187–4194, 2002.
67. A. V. Diniz, N. G. Ferreira, E. J. Corat, and V. J. Trava-Airoldi, "Efficiency study of perforated diamond electrodes for organic compounds oxidation process," Diamond and Related Materials, vol. 12, no. 3–7, pp. 577–582, 2003.
68. J.-F. Zhi, H.-B. Wang, T. Nakashima, T. N. Rao, and A. Fujishima, "Electrochemical incineration of organic pollutants on boron-doped diamond electrode, evidence for direct electrochemical oxidation pathway," The Journal of Physical Chemistry B, vol. 107, no. 48, pp. 13389–13395, 2003.
69. A. M. Polcaro, A. Vacca, S. Palmas, and M. Mascia, "Electrochemical treatment of wastewater containing phenolic compounds: oxidation at boron-doped diamond electrodes," Journal of Applied Electrochemistry, vol. 33, no. 10, pp. 885–892, 2003.
70. P. Cañizares, J. García-Gómez, C. Sáez, and M. A. Rodrigo, "Electrochemical oxidation of several chlorophenols on

diamond electrodes. Part I. Reaction mechanism," Journal of Applied Electrochemistry, vol. 33, no. 10, pp. 917–927, 2003.
71. P. Cañizares, J. García-Gómez, J. Lobato, and M. A. Rodrigo, "Electrochemical oxidation of aqueous carboxylic acid wastes using diamond thin-film electrodes," Industrial & Engineering Chemistry Research, vol. 42, no. 5, pp. 956–962, 2003.
72. A. Morão, A. Lopes, M. T. P. de Amorim, and I. C. Gonçalves, "Degradation of mixtures of phenols using boron doped diamond electrodes for wastewater treatment," Electrochimica Acta, vol. 49, no. 9-10, pp. 1587–1595, 2004.
73. P. Cañizares, C. Sáez, J. Lobato, and M. A. Rodrigo, "Electrochemical oxidation of polyhydroxybenzenes on boron-doped diamond anodes," Industrial & Engineering Chemistry Research, vol. 43, no. 21, pp. 6629–6637, 2004.
74. P. Cañizares, J. Lobato, R. Paz, M. A. Rodrigo, and C. Sáez, "Electrochemical oxidation of phenolic wastes with boron-doped diamond anodes," Water Research, vol. 39, no. 12, pp. 2687–2703, 2005.
75. C. Flox, J. A. Garrido, R. M. Rodríguez, et al., "Degradation of 4,6-dinitro-o-cresol from water by anodic oxidation with a boron-doped diamond electrode," Electrochimica Acta, vol. 50, no. 18, pp. 3685–3692, 2005.
76. M. J. Pacheco, A. Morão, A. Lopes, L. Ciríaco, and I. Gonçalves, "Degradation of phenols using boron-doped diamond electrodes: a method for quantifying the extent of combustion," Electrochimica Acta, vol. 53, no. 2, pp. 629–636, 2007.
77. M. Mascia, A. Vacca, S. Palmas, and A. M. Polcaro, "Kinetics of the electrochemical oxidation of organic compounds at BDD anodes: modelling of surface reactions," Journal of Applied Electrochemistry, vol. 37, no. 1, pp. 71–76, 2007.
78. X. Zhu, S. Shi, J. Wei et al., "Electrochemical oxidation characteristics of p-substituted phenols using a boron-doped diamond electrode," Environmental Science and Technology,

vol. 41, no. 18, pp. 6541–6546, 2007.
79. C. Flox, P.-L. Cabot, F. Centellas et al., "Solar photoelectro-Fenton degradation of cresols using a flow reactor with a boron-doped diamond anode," Applied Catalysis B: Environmental, vol. 75, no. 1-2, pp. 17–28, 2007.
80. M. Mascia, A. Vacca, A. M. Polcaro, S. Palmas, J. R. Ruiz, and A. da Pozzo, "Electrochemical treatment of phenolic waters in presence of chloride with boron-doped diamond (BDD) anodes: experimental study and mathematical model," Journal of Hazardous Materials, vol. 174, no. 1–3, pp. 314–322, 2010.
81. X. Zhu, J. Ni, H. Li, Y. Jiang, X. Xing, and A. G. L. Borthwick, "Effects of ultrasound on electrochemical oxidation mechanisms of p-substituted phenols at BDD and PbO_2 anodes," Electrochimica Acta, vol. 55, no. 20, pp. 5569–5575, 2010.
82. X. Zhu, J. Ni, J. Wei, X. Xing, H. Li, and Y. Jiang, "Scale-up of BDD anode system for electrochemical oxidation of phenol simulated wastewater in continuous mode," Journal of Hazardous Materials, vol. 184, no. 1–3, pp. 493–498, 2010.
83. J. Wei, X. Zhu, and J. Ni, "Electrochemical oxidation of phenol at boron-doped diamond electrode in pulse current mode," Electrochimica Acta, vol. 56, no. 15, pp. 5310–5315, 2011.
84. J. Sun, H. Lu, H. Lin et al., "Electrochemical oxidation of aqueous phenol at low concentration using Ti/BDD electrode," Separation and Purification Technology, vol. 88, pp. 116–120, 2012.
85. J. Lv, Y. Feng, J. Liu, Y. Qu, and F. Cui, "Comparison of electrocatalytic characterization of boron-doped diamond and SnO_2 electrodes," Applied Surface Science, vol. 283, pp. 900–905, 2013. ··
86. G. Hurwitz, P. Pornwongthong, S. Mahendra, and E. M. V. Hoek, "Degradation of phenol by synergistic chlorine-enhanced photo-assisted electrochemical oxidation," Chemical Engineering Journal, vol. 240, pp. 235–243, 2014.

87. X. Chen, G. Chen, F. Gao, and P. L. Yue, "High-performance Ti/BDD electrodes for pollutant oxidation," Environmental Science and Technology, vol. 37, no. 21, pp. 5021–5026, 2003.

88. X. Chen, F. Gao, and G. Chen, "Comparison of Ti/BDD and Ti/SnO$_2$-Sb$_2$O$_5$ electrodes for pollutant oxidation," Journal of Applied Electrochemistry, vol. 35, no. 2, pp. 185–191, 2005.

89. E. Weiss, K. Groenen-Serrano, and A. Savall, "A comparison of electrochemical degradation of phenol on boron doped diamond and lead dioxide anodes," Journal of Applied Electrochemistry, vol. 38, no. 3, pp. 329–337, 2008.

90. A. Medel, E. Bustos, K. Esquivel, L. A. Godínez, and Y. Meas, "Electrochemical incineration of phenolic compounds from the hydrocarbon industry using boron-doped diamond electrodes," International Journal of Photoenergy, vol. 2012, Article ID 681875, 6 pages, 2012.

91. J. H. Bezerra, M. M. Soares, N. Suely, D. Riveiro, and C. A. Martínez-Huitle, "Application of electrochemical oxidation as alternative treatment of produced water generated by Brazilian petrochemical industry," Fuel, vol. 96, pp. 80–87, 2012.

92. A. J. C. da Silva, E. V. dos Santos, C. C. D. O. Morais, C. A. Martínez-Huitle, and S. S. L. Castro, "Electrochemical treatment of fresh, brine and saline produced water generated by petrochemical industry using Ti/IrO$_2$-Ta$_2$O$_5$ and BDD in flow reactor," Chemical Engineering Journal, vol. 233, pp. 47–55, 2013.

93. R. B. A. Souza and L. A. M. Ruotolo, "Electrochemical treatment of oil refinery effluent using boron-doped diamond anodes," Journal of Environmental Chemical Engineering, vol. 1, no. 3, pp. 544–551, 2013.

94. A. Medel, E. Bustos, L. M. Apátiga, and Y. Meas, "Surface activation of C-sp^3 in Boron-Doped diamond electrode," Electrocatalysis, vol. 4, no. 4, pp. 189–195, 2013.

95. Mexican Regulation NMX-AA-050-SCFI-2001, "Analysis

of water-determination of total phenols in natural, potable, residual and treated residual waters," 2001.

96. A. Fakhru›l-Razi, A. Pendashteh, L. C. Abdullah, D. R. A. Biak, S. S. Madaeni, and Z. Z. Abidin, "Review of technologies for oil and gas produced water treatment," Journal of Hazardous Materials, vol. 170, no. 2-3, pp. 530–551, 2009.

97. L. Wei, S. Guo, G. Yan, C. Chen, and X. Jiang, "Electrochemical pretreatment of heavy oil refinery wastewater using a three-dimensional electrode reactor," Electrochimica Acta, vol. 55, no. 28, pp. 8615–8620, 2010.

98. E. V. dos Santos, S. F. M. Sena, D. R. da Silva, S. Ferro, A. de Battisti, and C. A. Martínez-Huitle, "Scale-up of electrochemical oxidation system for treatment of produced water generated by Brazilian petrochemical industry," Environmental Science and Pollution Research, vol. 21, pp. 8466–8475, 2014.

99. M. Murugananthan, S. S. Latha, G. B. Raju, and S. Yoshihara, "Role of electrolyte on anodic mineralization of atenolol at boron doped diamond and Pt electrodes," Separation and Purification Technology, vol. 79, no. 1, pp. 56–62, 2011.

100. A. Sánchez-Carretero, C. Sáez, P. Cañizares, and M. A. Rodrigo, "Electrochemical production of perchlorates using conductive diamond electrolyses," Chemical Engineering Journal, vol. 166, no. 2, pp. 710–714, 2011. ·

101. Official Mexican Regulation NOM-001-SEMARNAT-1996, "Maximum limits allowed of pollutants in discharges of wastewaters and national resources," 1996.

102. B. F. Giannetti, W. A. Moreira, S. H. Bonilla, C. M. V. B. Almeida, and T. Rabóczkay, "Towards the abatement of environmental mercury pollution: an electrochemical characterization," Colloids and Surfaces A: Physicochemical and Engineering Aspects, vol. 276, no. 1–3, pp. 213–220, 2006.

103. U. Skyllberg, P. R. Bloom, J. Qian, C.-M. Lin, and W. F. Bleam, "Complexation of mercury(II) in soil organic matter: EXAFS evidence for linear two-coordination with reduced sulfur

groups,"Environmental Science and Technology, vol. 40, no. 13, pp. 4174–4180, 2006.
104. D. T. Cestarolli and A. R. de Andrade, "Electrochemical oxidation of phenol at $Ti/Ru_{0.3}Pb_{(0.7-x)}Ti_xO_y$ electrodes in aqueous media," Journal of the Electrochemical Society, vol. 154, no. 2, pp. E25–E30, 2007.
105. S. Balaji, S. J. Chung, R. Thiruvenkatachari, and I. S. Moon, "Mediated electrochemical oxidation process: electro-oxidation of cerium(III) to cerium(IV) in nitric acid medium and a study on phenol degradation by cerium(IV) oxidant," Chemical Engineering Journal, vol. 126, no. 1, pp. 51–57, 2007.
106. D. Reyter, D. Bélanger, and L. Roué, "Nitrate removal by a paired electrolysis on copper and Ti/IrO_2 coupled electrodes—influence of the anode/cathode surface area ratio," Water Research, vol. 44, no. 6, pp. 1918–1926, 2010.
107. B. H. Diya›Uddeen, W. M. A. W. Daud, and A. R. Abdul Aziz, "Treatment technologies for petroleum refinery effluents: a review," Process Safety and Environmental Protection, vol. 89, no. 2, pp. 95–105, 2011.

Chapter 2

Extraction of S- and N-compounds from the Mixture of Hydrocarbons by Ionic Liquids as Selective Solvents

Beata Gabrić[1], Aleksandra Sander[2],
Marina Cvjetko Bubalo[3], and Dejan Macut[2]

[1]INA-Industrija Nafte d.d., Avenija Veceslava Holjevca 10, 10000 Zagreb, Croatia

[2]Faculty of Chemical Engineering and Technology, University of Zagreb, Marulićev Trg 19, 10000 Zagreb, Croatia

[3]Faculty of Food Technology and Biotechnology, University of Zagreb, Pierottijeva 6, 10000 Zagreb, Croatia

ABSTRACT

Liquid-liquid extraction is an alternative method that can be used for desulfurization and denitrification of gasoline and diesel fuels. Recent approaches employ different ionic liquids as selective solvents, due to their general immiscibility with gasoline and diesel, negligible vapor pressure, and high selectivity to sulfur- and nitrogen-containing compounds. For that reason, five imidazolium-based ionic liquids and one pyridinium-based ionic liquid were selected for extraction of thiophene, dibenzothiophene, and pyridine from two model solutions. The influences of hydrodynamic conditions, mass ratio, and number of stages were investigated. Increasing the mass ratio of ionic liquid/model fuel and multistage extraction promotes the desulfurization and denitrification abilities of the examined ionic liquids. All selected ionic liquids can be reused and regenerated by means of vacuum evaporation.

INTRODUCTION

Diesel and gasoline rich in sulfur will produce exhaust gases containing SO_x, a major contributor to air pollution and acid rain. Due to strict regulatory requirements on sulfur content in fuels, this problem is widely investigated [1–15].

Today, a commonly used industrial process for desulfurization of diesel and gasoline is hydrodesulphurization (HDS), in which organic sulfur compounds are converted to H_2S and the corresponding hydrocarbons. The process requires high temperatures and pressures, as well as large amounts of hydrogen, making HDS an expensive and relatively environmentally unfriendly process. Moreover, this process requires even more expensive process conditions for removal of compounds such as dibenzothiophene derivatives [1–4, 7]. For that reason, scientists explore new environmentally friendly and energy-saving processes [1, 5].

Liquid-liquid extraction is one of such processes. The major advantage of extraction is in its mild operating conditions (low temperatures and pressures—ambient conditions) and consequently conservation of significant amount of energy. As for any other process that involves mass-separating agent, special care must be taken when selecting a proper solvent [16]. However, liquid-liquid extraction generates organic waste which requires disposal. In order to minimize that waste, the solvent should be recyclable, reusable, and regenerable [4–6, 13, 14, 16, 17]. Solvents commonly used in industry are volatile organic compounds that exhibit high-vapor pressure. Additionally, volatile organic solvents are flammable and are of varying toxicity, depending on their nature, so the major task of many researchers is to find a way of replacing VOCs with environmentally friendly, so-called green solvents [18]. This is not a simple problem, because one cannot simply replace one solvent with another. One possible solution is replacement of organic solvents with ionic liquids. Ionic liquids are very good solvents for a wide range of organic, inorganic, and polymeric compounds. Owing to their negligible vapor pressure, ionic liquids are considered as green solvents. But even if ionic liquids are not responsible for air pollution, their impact on complete ecosystem is still unexplored. For instance, many ionic liquids are soluble in water, so they can enter the environment by this path [1].

During the past decade, an increasing trend of investigations concerning the possibilities of replacing the existing desulfurisation processes with the extraction of fuels by ecologically acceptable ionic liquids can be observed. Authors mostly use model solutions of different compositions as the representatives of real diesel and gasoline [3, 5, 6, 8–11, 13, 15, 16] as well as real refinery samples [2, 4, 7, 8, 15, 16]. Most commonly used are the imidazolium-based ionic liquids with 1-alkyl-3-methylimidazolium cations ([CnC1Im]) and various anions. The use of the pyridinium-based ionic liquids for desulfurization of diesel and gasoline was also investigated [7, 14, 16]. Denitrogenation of feeds before HDS was also investigated in order to increase the efficiency of HDS process because N-compounds compete with S-compounds on the catalyst surface [19]. Pyridinium and imidazolium ionic liquids were tested, and it was observed that the same ionic liquids have extraction ability to both S- and N-compounds [1, 20, 21].

The aim of this paper is to investigate the applicability of six different ionic liquids for the separation of sulfur and nitrogen compounds from model diesel and FCC gasoline by means of liquid-liquid extraction. The time needed to reach equilibrium in a batch extractor equipped with the mechanical stirrer was also determined. The influences of the hydrodynamic conditions, the mass ratio (ionic liquid/model solution), and the number of extraction stages on the extraction effectiveness were investigated. In addition, the reusability of the selected ionic liquids and regeneration by means of vacuum evaporation was tested.

MATERIALS AND METHODS

Chemicals

The list of chemicals used for preparation of model solutions as representatives of FCC gasoline and diesel is as follows: n-hexane, n-heptane, toluene, and pyridine were purchased from Carlo Erba Reagenti; thiophene, n-hexadecane, and dibenzothiophene were purchased from Acros Organics; n-dodecane was purchased from Fisher UK and isooctane from Kemika.

The list of chemicals used in the synthesis of ionic liquids is as follows: 3,5-lutidine, 1-bromohexane, and lithium bis(trifluoromethanesulfon)imide were purchased from Acros Organics; 1-methylimidazole was purchased from Alfa Aesar; 1,2-dimethylimidazole, 1-bromopentane, 1-bromoheptane, 1-bromodecane, and benzyl chloride were purchased from Sigma-Aldrich; acetonitrile was purchased from J.T. Baker, ethyl acetate and toluene from Kemika and dichloro methane from TTT Ltd.

1-Ethyl-3-methylimidazolium ethylsulfate was purchased from MERCK.

Model Fuels

Composition of model fuels is presented in Table 1. Representatives of FCC gasoline and diesel are prepared according to the literature [21].

Table 1: Composition of the model fuels

Model 1 (FCC gasoline)	Model 2 (diesel)
26% n-hexane	3% thiophene
6% thiophene	26% heptanes
26% isooctane	6% pyridine
26% n-heptane	10% toluene
6% pyridine	26% n-dodecane
10% toluene	26% n-hexadecane
	3% dibenzothiophene

General Procedure for the Preparation of Ionic Liquids

Quaternization of heterocyclic imidazolium or pyridinium compound followed by the anion metathesis was performed according to standard procedures [22–24] or their modifications. Aliphatic halide (1-bromopentane, 1-bromohexane, 1-bromoheptane, or 1-bromodecane) was added in 10% excess to the stirred solution of heterocyclic imidazolium or pyridinium compound (1-methylimidazole, 1,2-dimethylimidazole, or 3,5-dimethylpyridine) in toluene at 0°C, and the reaction mixture was stirred 24–48 h at 70°C. Sedimented quaternary imidazolium or pyridinium halide was washed thoroughly with ethyl-acetate and dried under reduced pressure. The anion metathesis was performed by the treatment of the aqueous solution of obtained halides with lithium bis(trifluoromethanesulfonyl)imide (LiTf$_2$N) in excess of 10% and was stirred for approximately 2 h. The upper aqueous phase was decanted, and the lower product portion was washed with water until chloride free, as determined by silver nitrate test. All ILs were dried under high vacuum at 90°C for 8 h prior to use and were characterized by ^{1}H NMR.

^{1}H NMR spectra were recorded in DMSO-d$_6$ on a Bruker AV300 (300 MHz) spectrometer at Ruđer Bošković Institute (Zagreb, Croatia). Chemical shifts were expressed in ppm values using TMS as an internal standard.

The selected ionic liquids are presented in Table 2.

Table 2: Structures of selected ionic liquids

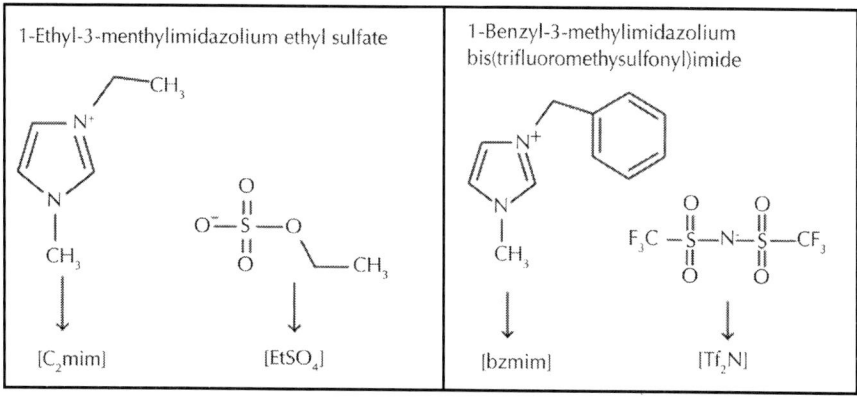

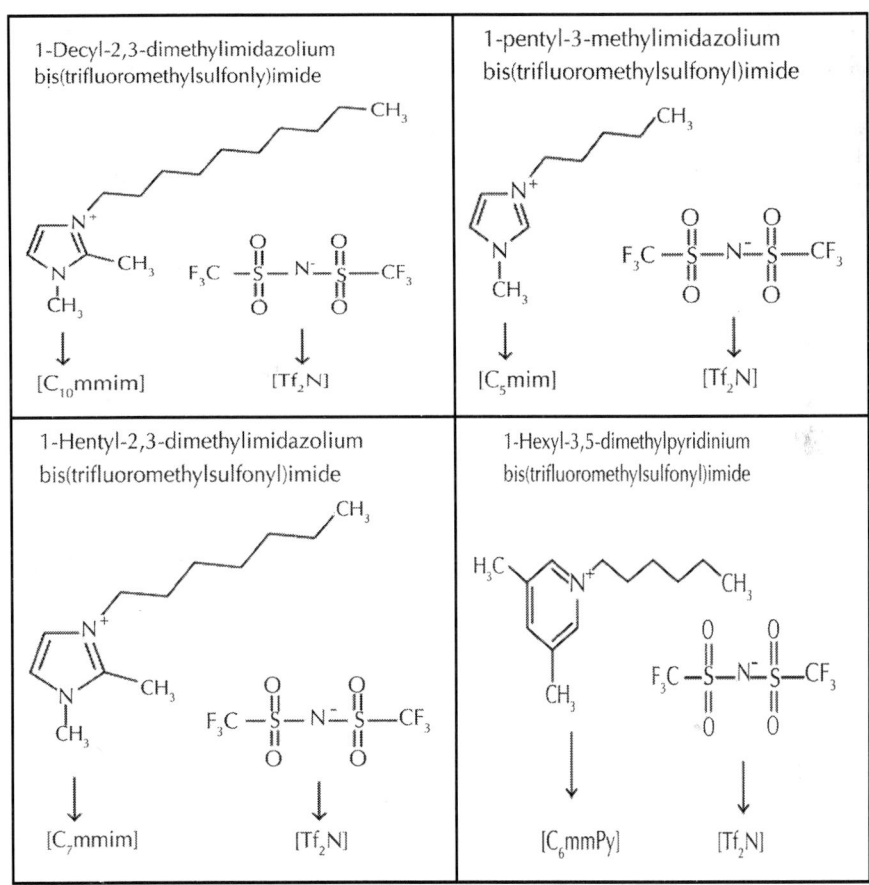

Measurement of Density, Viscosity, Surface Tension, and Interfacial Tension

Viscosity of ionic liquids and model solutions were measured on the Brookfield DV-III Ultra Programmable Rheometer.

Density, surface tension, and interfacial tension were measured on the KRUSS K9 tensiometer.

All measurements were performed at the room conditions (25°C).

Liquid-liquid Extraction

Liquid-liquid extraction experiments were carried out in a laboratory batch extractor (mixing vessel) equipped with the mechanical stirrer. Quantities of solvent and feed solution were mixed together at the defined mass ratio (ionic liquid/model solution) for a given period of time. The mixing rate was set to the value at which complete dispersion is achieved. This state was visually observed. All systems were mixed with the mixing rate of 500 rpm. Afterward, the heavier and lighter phases are separated in a settling unit.

Experiments have been performed at four mass ratios of ionic liquid to model solutions (0.25, 0.50, 0.75, and 1.00).

Mass ratio of ionic liquid and model solution was calculated by

$$S = \frac{m(\text{ionic liquid})}{m(\text{model solution})}. \tag{1}$$

Extraction efficiency was calculated by the following equation:

$$\varepsilon = \frac{x_F - x_R}{x_F}. \tag{2}$$

The influence of mixing intensity (100–500 rpm) and the number of stages were also investigated. These experiments were performed at the lowest mass ratio (ionic liquid/model solution), S=0.25

Gas Chromatography

Concentration of all components was determined by means of gas chromatography. The GC used was a GC-2014-Shimadzu equipped with an autosampler, FID detector, and fused silica capillary column CBP1-S25-050 (length: 25 m, inner diameter: 0.32 mm). The GC program parameters for the analyses of both model solutions are presented in the Supplementary Material

Calibration curves were measured and incorporated in the software (Shimadzu GC Solutions), so after analysis mass fractions of all components were calculated automatically.

Regeneration of Ionic Liquids

Due to the negligible vapor pressure of ionic liquids and volatility of components present in ionic liquid after extraction, as a recovery method a vacuum evaporation was selected. For that purpose, an IKA rotary evaporator equipped with the vacuum pump was used. All ILs were dried under high vacuum at 80°C for 8 h. After evaporation, the purity of ionic liquids was determined by means of ^{1}H NMR spectroscopy.

RESULTS AND DISCUSSION

The aim of this paper was to select appropriate ionic liquids for desulfurization and denitrification of model gasoline and diesel fuel. As previously mentioned, five of the selected ionic liquids have imidazolium-based cation, and the last one has pyridinium-based cation. Imidazolium-based ionic liquids mostly have the same anion, bis(trifluoromethylsulfonyl)imide. 1-Ethyl-3-methyl imidazolium ethyl sulfate was selected from the literature since, according to the literature cited [21], this ionic liquid is immiscible with the prepared model solutions and real samples. The same statement holds for the pyridinium-based ionic liquid, 1-hexil-3-methylpyridinium bis(trifluoromethylsulfonyl)imide. All ionic liquids were synthesized, except the commercially available [C_2mim][EtSO$_4$].

Physical Properties of Model Solutions and Ionic Liquids

Besides the distribution coefficient, selectivity, mutual solubility of solvents, regeneration, and physical properties of both phases (such as density, viscosity, and interfacial tension) will influence the selection of the most appropriate solvent for liquid-liquid extraction. For the investigated model solutions and ionic liquids, measured results were presented in Table 3. It can be seen that differences in properties between the model solutions and ionic liquids were satisfactory. From the viscosity point of view, the most suitable ionic liquid is $[C_5mim][Tf_2N]$ due to the lowest viscosity. The same ionic liquid should be the most appropriate from the interfacial tension point of view. Liquid with lower viscosity and interfacial tension will be easily dispersed into the other, and higher rates for mass transfer will be achieved. The reader should have in mind that the final conclusion about the rate of mass transfer, the solubility of all components of the model solution, the distribution coefficients, and, selectivity must be determined.

Table 3: Physical properties of model solutions and ionic liquids

	ρ, kg m^{-3}	η, Pa s	σ, mN m^{-1}	σ_{1-2}, mN m^{-1}	
				Model 1	Model 2
Model 1	720	4.56·10^{-4}	21.3		
Model 2	741	1.73·10^{-3}	24.9		
$[C_2mim][EtSO_4]$	1236	0.0896	49.3	10.6	12.3
$[C_5mim][Tf_2N]$	1404	0.0525	32.6	3.2	5.6
$[C_6mmPy][Tf_2N]$	1332	0.1152	34.7	3.8	9.7
$[bzmim][Tf_2N]$	1491	0.1508	41.5	6.0	6.2
$[C_7mmim][Tf_2N]$	1226	0.1048	46.9	10.5	12.1
$[C_{10}mmim][Tf_2N]$	1269	0.1472	33.3	3.4	5.8

Determination of Extraction Time

In order to determine the time needed for maximum extraction efficiency, extraction experiments were stopped in defined time intervals. Experiments were performed at the mass ratio of ionic liquid and model solution equal to 0.25. After 10 minutes during which the phases were separated, very small amount of model solution was taken, and the concentration was measured. The procedure was taken from the literature data [2]. With each model solution, experiments have been performed with all ionic liquids.

Results obtained after extraction with [C$_5$mim][Tf$_2$N] for both model solutions are presented in Figure 1. For all key components (thiophene, pyridine, and dibenzothiophene), the maximum extraction efficiency was achieved after 20 minutes. All other ionic liquids act similarly, so extraction time was set to 20 minutes.

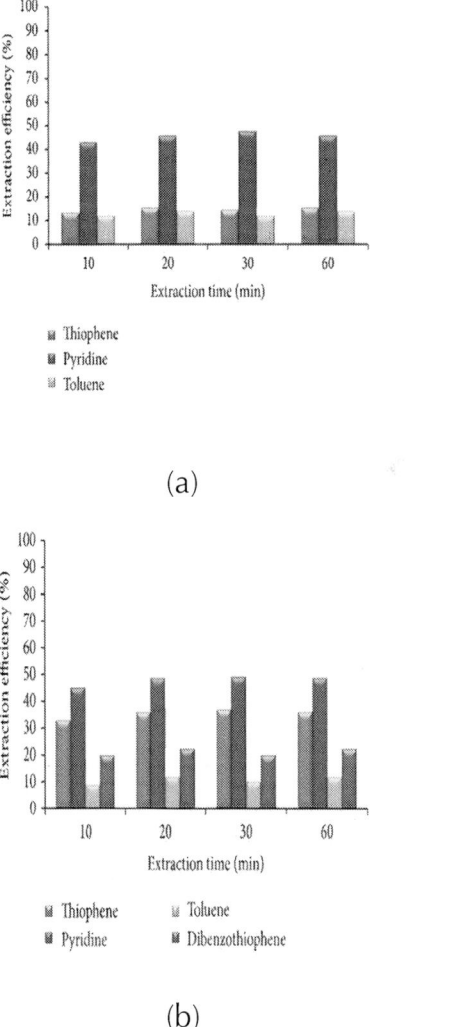

(a)

(b)

Figure 1: The influence of extraction time on the extraction efficiency ([C$_5$mim][Tf$_2$N];S=0,25).

The Influence of the Hydrodynamic Conditions

In order to see whether the hydrodynamic conditions influence the mass transfer during extraction of S- and N-compounds from the mixture of hydrocarbons with the selected ionic liquids, the experiments have been performed at different mixing intensities. It is well known that higher mixing intensities will result in enhanced mass transfer. At higher mixing rate, resistances to mass transfer are lower, and larger mass transfer area is produced (smaller droplets). The mixing rate was changed from 100 to 500 rpm. The Reynolds number ranges from 1705 for [C_7mmim][Tf_2N] (at 100 rpm) up to 9594 for [bzmim][Tf_2N] (at 500 rpm), for model solution that represents FCC gasoline. For model solution that represents diesel fuel, the Reynolds number ranges from 536 for [C_7mmim][Tf_2N] (at 100 rpm) up to 3030 for [bzmim][Tf_2N] (at 500 rpm). Based on the obtained results, it can be stated that the hydrodynamic conditions do not influence the extraction efficiency of the selected ionic liquids (Figure 2). Physical properties of the involved phases and solubility of key components play a huge role in the transfer of key components between the model solution and selected ionic liquids. The rate of mass transfer is very high, so equilibrium was achieved after short time (Figure 1), even at the lowest mass transfer area (at 100 rpm). High rates of different processes and chemical reactions were reported in the literature [25], as one of the advantages of ionic liquids. Another advantage that is confirmed by our results is high solubility of S- and N-compounds in ionic liquids. For all other ionic liquids and model solutions, the maximum extraction efficiency was also obtained at the lowest mixing rate.

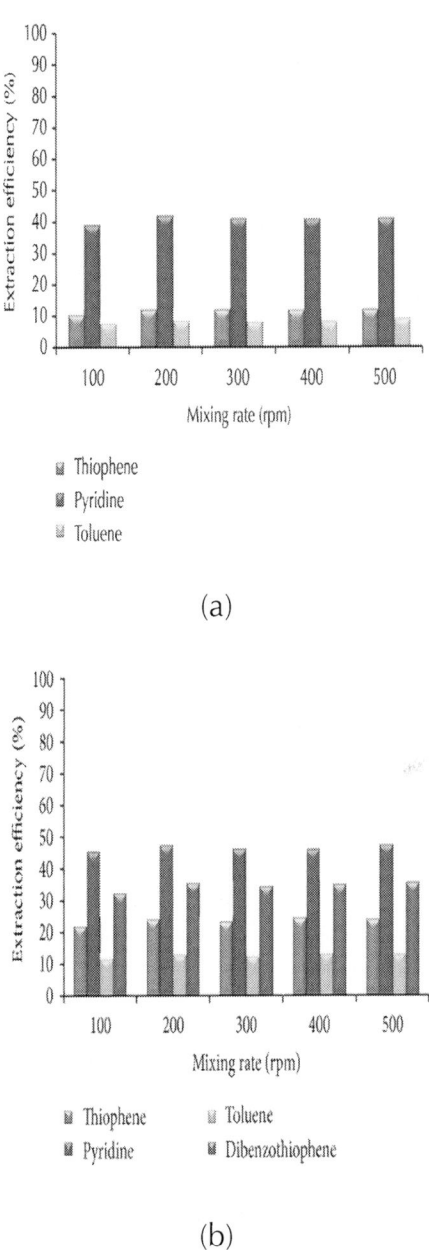

Figure 2: The influence of the mixing intensity on the extraction efficiency ([C_6mmPy][Tf_2N];S=0,25 ;t=20 min): (a) model solution 1; (b) model solution 2.

Initial Screening of Ionic Liquids

With the purpose of initial screening, experiments were conducted at the mass ratio of ionic liquids and model solutions equal to 0.25. After phase separation, the compositions of model solutions were determined by means of gas chromatography. All investigated ionic liquids satisfy the major requirement during selection of solvent. In other words, all of them are insoluble in model solutions. These results are in concordance with the previously reported data [7] obtained with the same model solutions when [C_2mim] [$EtSO_4$] and [C_6mmPy] [Tf_2N] were used as selective solvents.

For the model solution that represents the FCC gasoline (model solution 1), results are presented in Figure3 (a). Since one of the most important conditions that solvent must fulfill is selectivity, other components of model solution should not be extracted with the selected ionic liquid. n- Hexane, n-isooctane, and n-heptane are insoluble in all investigated ionic liquids. On the other hand, toluene, as a representative of the aromatic hydrocarbons in model solution, is soluble in all ionic liquids except in [C_2mim] [$EtSO_4$]. Partial dearomatization is favorable [21], so this ionic liquid is the least suitable ionic liquid. Basically, extraction efficiency of all ionic liquids is higher for pyridine than for thiophene. The influence of the length of the alkyl chain can be observed for the poly substituted imidazolium-based ionic liquids with long alky chain ([C_7mmim][Tf_2N] and [C_{10}mmim][Tf_2N]). Higher alkyl chain results in slightly higher extraction efficiency. Obtained result is in concordance with results obtained by other researchers that investigated the influence of the length of alkyl chain on the desulfurization efficiency of extraction of thiophene and pyridine from model fuels [2, 13, 14]. Extraction efficiency of all investigated ionic liquids for thiophene has a similar numerical value. Based on the obtained results, for denitrification and desulfurization of this model solution, the most suitable ionic liquid is [C_5mim][Tf_2N]. This ionic liquid has the lowest interfacial tension and viscosity (Table 3), so favorable conditions for mass transfer were achieved.

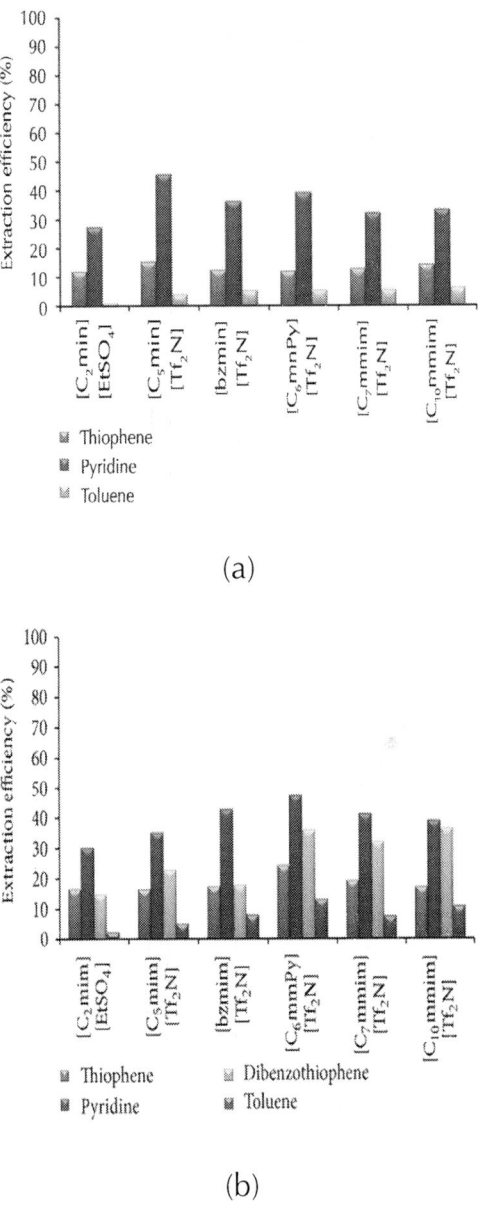

Figure 3: Initial screening of ionic liquids: (a) model solution 1; (b) model solution 2.

For the model solution that represents the diesel fuel (model solution 2), extraction efficiency of all investigated ionic liquids is shown in Figure 3(b). n-Heptane, n-dodecane, and n-hexadecane are insoluble in all investigated ionic liquids, so requirement concerning the selectivity was fulfilled. The majority of the investigated ionic liquids have higher affinity to thiophene and pyridine in the model diesel solution in comparison to the results obtained with gasoline model solution. Zhang et al. [9] observed quite opposite results with a set of different ionic liquids. Obviously, cation structure, selected anion, and its mutual interaction as well as interactions between compounds of the feed and the charged ion pairs of the ionic liquid influence the extraction capacity of ionic liquid. The extraction of dibenzothiophene is more difficult than extraction of thiophene and pyridine due to its complex structure, so this should be the major criteria for the selection of the most suitable ionic liquid. Having in mind that partial dearomatization is favorable, and ionic liquid selectivity to the key components (S- and N-compounds) $[C_6mmPy][Tf_2N]$ can be the best choice. This ionic liquid shows the highest extraction efficiency to all key components. Poly substituted ionic liquids with long alkyl chain show high extraction capacity to dibenzothiophene. The usage of these two ionic liquids ($[C_7mmim][Tf_2N]$ and $[C_{10}mmim][Tf_2N]$) should be further investigated from the ecological and toxicological points of view, because it was observed that toxicity is higher for longer alkyl chain substituents [25].

With this model solution, no general conclusions concerning the physical properties of the ionic liquids can be drawn. Different results can be found in the literature. For instance, Cern et al. [20] have reported that lower-viscosity ionic liquid is more effective for denitrogenation of real diesel. The authors have discussed their results also with different types of substituents on the pyridinium and imidazolium rings. Wilfred et al. [5] have investigated the influence of different process conditions and types of ionic liquids on the extraction efficiency of dibenzothiophene from dodecane. Based on their results, extraction efficiency increases with decrease of ionic liquid density. Both statements hold, but as previously mentioned many factors influence the mass transfer, not only physical properties.

The Influence of (Ionic Liquid/Model Solution) Mass Ratio

The influence of the mass ratio of ionic liquid to model solution on the extraction efficiency is presented in Figures 4 and 5. For both model solutions, results obtained with $[C_5mim][Tf_2N]$ and $[C_6mmPy][Tf_2N]$ are presented graphically, while the other results are commented.

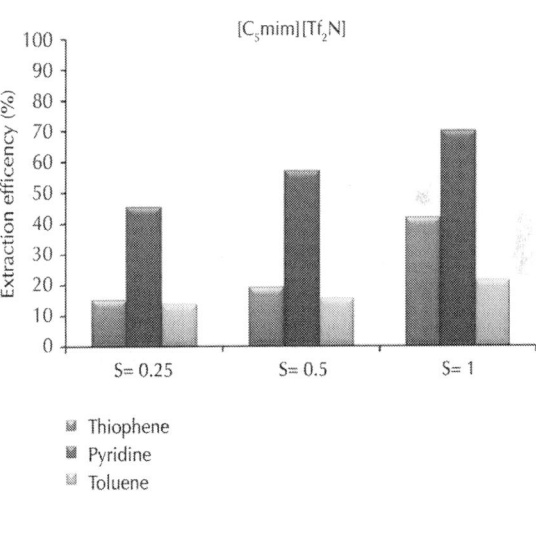

(a)

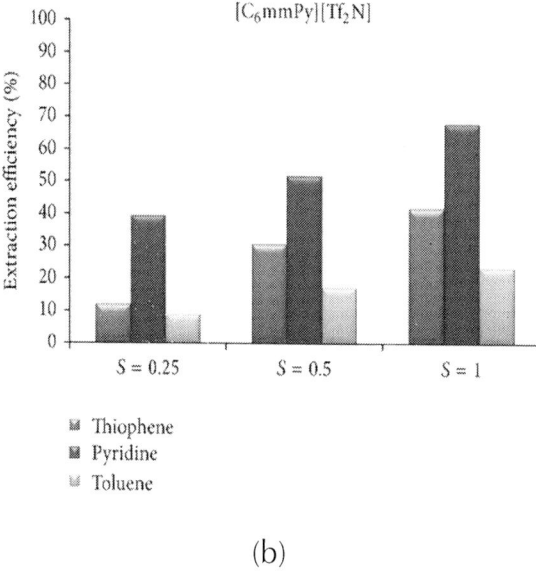

(b)

Figure 4: The influence of the mass ratio (ionic liquid/model solution) on the extraction efficiency of selected ionic liquids.

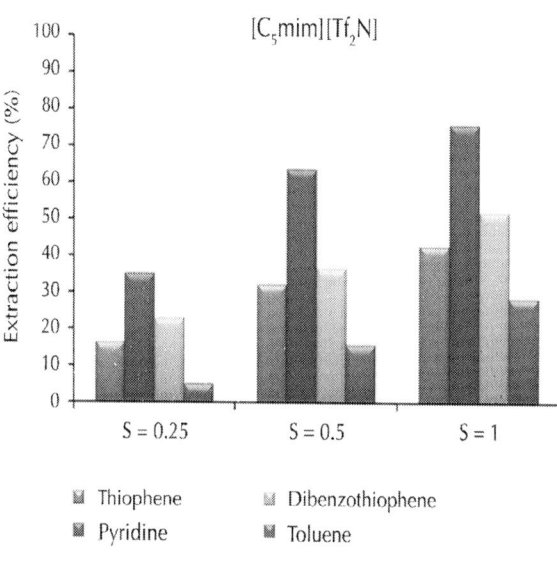

(a)

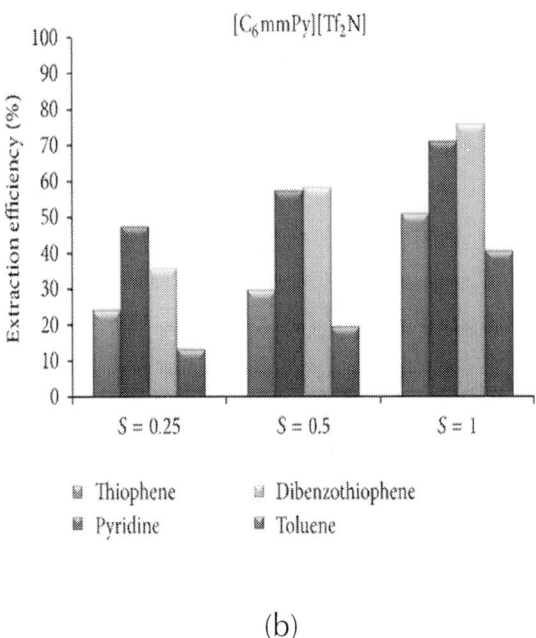

(b)

Figure 5: The influence of the mass ratio (ionic liquid/model solution) on the extraction efficiency of selected ionic liquids.

System Model FCC-Ionic Liquid

Higher mass ratio results in more efficient extraction of all soluble components. This influence was more pronounced for thiophene and pyridine than for toluene. For the system model FCC-[C_2mim][EtSO$_4$], extraction efficiency to thiophene was increased up to 26.7% and to pyridine up to 41.0%, when mass ratio, S, was increased from 0.25 to 1.00. For the system model FCC-[bzmim][Tf$_2$N], extraction efficiency to thiophene was increased up to 17.7% and to pyridine up to 53.3%. When [C_7mmim][Tf$_2$N] was used, extraction efficiency to thiophene was increased up to 21.4% and to pyridine up to 50.8%. From the financial point of view, small mass ratio can be justified. Higher mass ratio means higher quantity of ionic liquid during extraction and finally higher amount of energy used for regeneration.

System Model Diesel-Ionic Liquid

For the system model diesel-ionic liquid, similar results were obtained. Higher mass ratio positively influences extraction efficiency of all ionic liquids for the key components (Figure 5). Extraction efficiency at mass ratio, S=1.0 for all other ionic liquids, is as follows: [C_2mim][EtSO_4] (thiophene: 40.3%; pyridine: 56.9%; dibenzothiophene: 37.8%); [bzmim][Tf_2N] (thiophene: 43.9%; pyridine: 73.8%; dibenzothiophene: 48.3%); [C_7mmim][Tf_2N] (thiophene: 21.4%; pyridine: 50.8%; dibenzothiophene: 33.7%); [C_{10}mmim][Tf_2N] (thiophene: 50.8%; pyridine: 78.8%; dibenzothiophene: 67.1%).

Regeneration

According to the literature, ionic liquids can be easily regenerated [5, 13–15, 18] and even used several times without regeneration [5]. For that reason, extraction was performed with contaminated and regenerated ionic liquids at the lowest mass ratio and previously determined mixing time. Obtained results for two selected ionic liquids are presented in Figures 6 and 8.

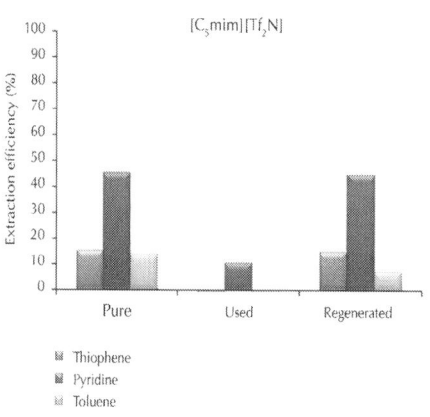

(a)

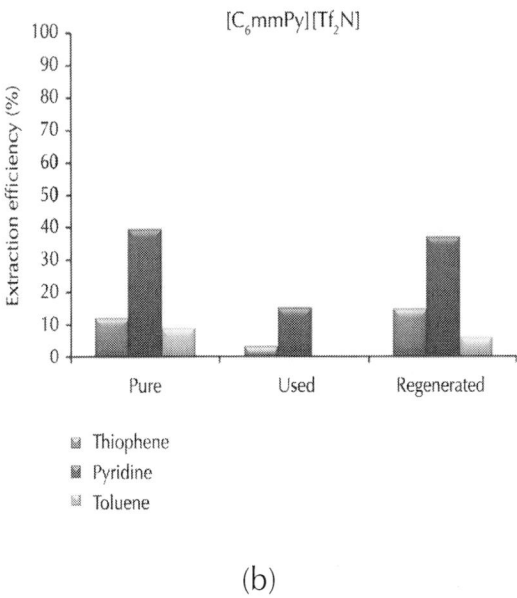

(b)

Figure 6: The influence of the solvent purity on the extraction efficiency (FCC model solution).

One of the most important properties of ionic liquids, its nonvolatility, was used for regeneration. After extraction, contaminated ionic liquids were first used for another experiment and then purified by means of vacuum evaporation. For the model solution that represents FCC gasoline, obtained results are shown in Figure 6. When contaminated ionic liquid was used, extraction efficiency was reduced. Thiophene and toluene were not transferred to the ionic liquid, probably due to the saturation of ionic liquid with these two components during first usage. After regeneration, ionic liquids were totally purified, since reduction of the extraction efficiency was not observed. ^{1}H NMR spectra (Figure 7) are the same for both fresh and regenerated ionic liquids. Peaks that belong to pyridine, thiophene, and toluene are visible on the used (contaminated) ionic liquid spectra.

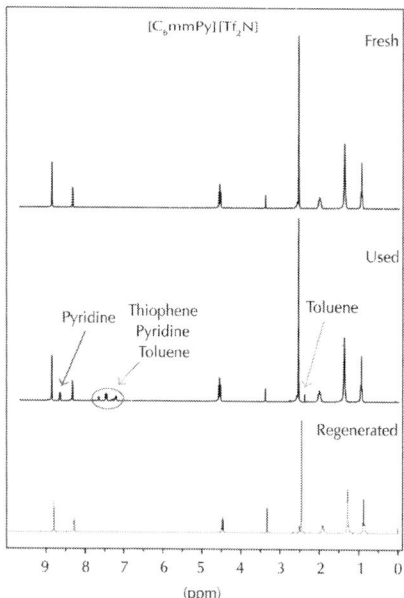

Figure 7: ^{1}H NMR spectra of fresh and used ionic liquids with model FCC solution and regenerated ionic liquids [C$_6$mmPy][Tf$_2$N].

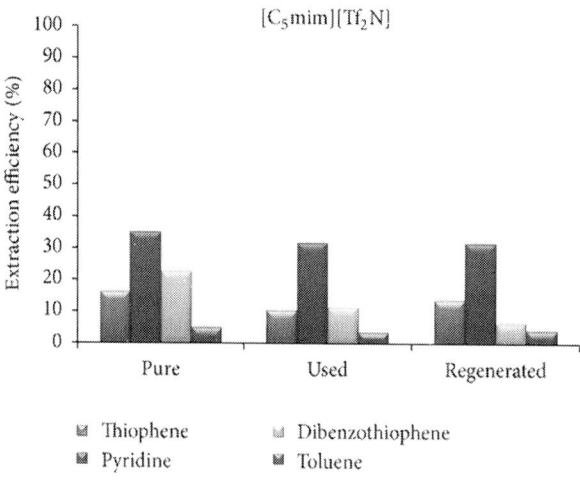

(a)

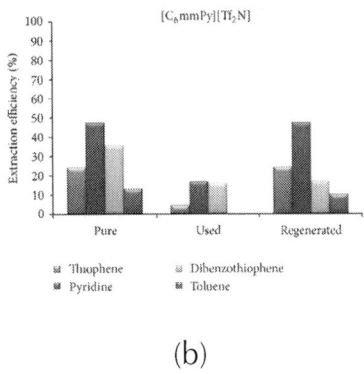

(b)

Figure 8: The influence of the solvent purity on the extraction efficiency (diesel model solution).

Extraction efficiency of the regenerated ionic liquids for desulfurization and denitrification of model diesel solution was decreased (Figure 8). In a way, this result was expected since the boiling point of dibenzothiophene is very high (332°C). Therefore, dibenzothiophene cannot be removed by vacuum evaporation, so extraction capacity of ionic liquids is reduced. ^{1}H NMR spectra of fresh, contaminated, and regenerated [C_5mim][Tf_2N] are shown in Figure 9.

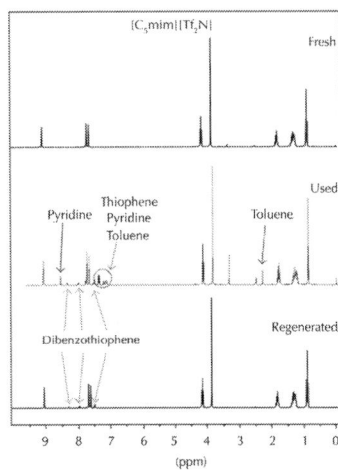

Figure 9: ^{1}H NMR spectra of fresh and used ionic liquids with model diesel solution and regenerated ionic liquids [C_5mim][Tf_2N].

Dibenzothiophene can be observed on the regenerated ionic liquid spectra. Dibenzothiophene can be removed by extraction with some suitable solvent, but in that way one additional regenerating step must be involved. As a consequence, the process will become economically and ecologically adverse. If ionic liquid is reused without regeneration, extraction efficiency for dibenzothiophene is drastically reduced, probably due to the saturation of ionic liquid with dibenzothiophene. This is an unwanted effect, because dibenzothiophene presents the major problem in most desulfurization processes [26].

Multistage Extraction

Multistage extraction has been performed with selected solvents for model FCC and model diesel solutions. Obtained results are presented in Table 4. Experiments have been performed at the lowest examined mass-ratio (S=0.25) and short mixing time (t=20min). Four-stage extractions were performed, with 20-minute interval of phase separation between each stage. As expected, by increasing the number of extraction stages, mass fractions of thiophene and pyridine in the model FCC solution were decreased.

Table 4: Extraction efficiency for multistage extraction of S- and N-compounds from the model solutions

Ionic liquid	Component	Extraction efficiency %							
		FCC model solution				Diesel model solution			
		stage				stage			
		1	2	3	4	1	2	3	4
[C$_2$mim] [EtSO$_4$]	Tiophene	12.01	24.11	32.98	39.85	16.60	25.66	39.94	49.36
	Pyridine	27.52	49.13	70.58	70.58	30.22	45.42	66.16	86.22
	DBT	—	—	—	—	14.70	25.14	33.32	42.40
[C$_2$mim] [Tf$_2$N]	Tiophene	15.37	26.36	35.93	45.92	16.41	32.44	45.82	53.58
	Pyridine	45.62	58.63	70.18	78.59	35.12	49.58	76.93	87.80
	DBT	—	—	—	—	22.78	41.98	54.42	66.29

[C$_6$mmPy] [Tf$_2$N]	Tiophene	12.05	24.71	40.06	50.53	24.21	29.53	43.08	65.59
	Pyridine	39.21	45.82	65.76	75.69	47.43	58.74	71.59	98.87
	DBT	—	—	—	—	35.67	59.10	71.38	92.61
[bzmim] [Tf$_2$N]	Tiophene	12.44	25.06	34.34	39.45	17.28	30.98	39.89	51.05
	Pyridine	36.19	54.72	68.37	70.17	42.82	62.59	72.59	80.57
	DBT	—	—	—	—	17.71	36.44	45.83	54.31
[C$_7$mmim] [Tf$_2$N]	Tiophene	12.73	26.15	35.29	40.45	19.27	32.19	37.06	45.87
	Pyridine	32.16	53.27	67.60	70.64	41.08	65.57	78.67	81.04
	DBT	—	—	—	—	31.89	53.74	59.08	70.37
[C$_{10}$mmim] [Tf$_2$N]	Tiophene	17.16	29.67	40.95	50.49	17.16	31.82	45.93	57.48
	Pyridine	32.48	54.36	66.98	77.11	38.85	60.43	74.08	81.51
	DBT	—	—	—	—	36.27	56.36	69.72	77.19

For the second group of the investigated system, model diesel-ionic liquid, extraction efficiency for all three key components is increased with increasing number of extraction stages. The highest efficiency was obtained with [C$_6$mmPy][Tf$_2$N] as the selective solvent.

If results obtained with the highest investigated mass ratio (S=1) are compared with results obtained after four stages, when the same amount of ionic liquid was used, it can be concluded that multistage extraction is more efficient. In multistage extraction, after each stage fresh ionic liquid was used and mixed with the feed with lower concentrations of the key components. In that way, the driving force for mass transfer was increased, so higher amount of key component was transferred between phases.

CONCLUSIONS

Among six tested ionic liquids, [C$_6$mmPy][Tf$_2$N] is found to be the most effective selective solvent for desulfurizationa and denitrification of model solution that represents diesel at room temperature. For the other investigated model solution, [C$_5$mim][Tf$_2$N] is the most appropriate selective solvent. The extraction should be carried out at the lower mass ratio, to reduce amount of ionic liquid and consequently reduce energy used for regeneration.

Ionic liquids can be successfully regenerated by means of vacuum evaporation. Regeneration was incomplete if model solution contains dibenzothiophene. If very low concentration of S- and N-compounds must be achieved, multistage extraction is the preferred choice.

It is necessary to reach an agreement between the defined purity and quality of gasoline and diesel fuel, by careful selection of the process conditions having in mind ecological and economical issues.

Experiments with real FCC gasoline and diesel fuel should be performed in order to see whether the selected ionic liquid can be used for desulfurization and denitrification. The results with real samples cannot be predicted because real fuels consist of large number of compounds that can influence extraction efficiency and mutual solubility of ionic liquid and fuel.

ACKNOWLEDGMENTS

The authors would like to thank INA-Industrija Nafte d.d. and the Croatian Academy of Sciences and Arts for the financial support.

REFERENCES

1. C. Song, "An overview of new approaches to deep desulfurization for ultra-clean gasoline, diesel fuel and jet fuel," Catalysis Today, vol. 86, no. 1–4, pp. 211–263, 2003.
2. X. Chu, Y. Hu, J. Li et al., "Desulfurization of diesel fuel by extraction with [BF_4]$^-$-based ionic liquids," Chinese Journal of Chemical Engineering, vol. 16, no. 6, pp. 881–884, 2008.
3. L. Alonso, A. Arce, M. Francisco, and A. Soto, "Thiophene separation from aliphatic hydrocarbons using the 1-ethyl-3-methylimidazolium ethylsulfate ionic liquid," Fluid Phase Equilibria, vol. 270, no. 1-2, pp. 97–102, 2008.
4. X. Jiang, Y. Nie, C. Li, and Z. Wang, "Imidazolium-based

alkylphosphate ionic liquids—a potential solvent for extractive desulfurization of fuel," Fuel, vol. 87, no. 1, pp. 79–84, 2008.

5. C. D. Wilfred, C. F. Kiat, Z. Man, M. A. Bustam, M. I. M. Mutalib, and C. Z. Phak, "Extraction of dibenzothiophene from dodecane using ionic liquids," Fuel Processing Technology, vol. 93, no. 1, pp. 85–89, 2012.

6. L. Alonso, A. Arce, M. Francisco, and A. Soto, "Solvent extraction of thiophene from n-alkanes (C_7, C_{12}, and C_{16}) using the ionic liquid [C_8mim][BF_4]," Journal of Chemical Thermodynamics, vol. 40, no. 6, pp. 966–972, 2008.

7. M. Francisco, A. Arce, and A. Soto, "Ionic liquids on desulfurization of fuel oils," Fluid Phase Equilibria, vol. 294, no. 1-2, pp. 39–48, 2010.

8. S. Zhang and Z. C. Zhang, "Novel properties of ionic liquids in selective sulfur removal from fuels at room temperature," Green Chemistry, vol. 4, no. 4, pp. 376–379, 2002.

9. S. Zhang, Q. Zhang, and Z. C. Zhang, "Extractive desulfurization and denitrogenation of fuels using ionic liquids," Industrial and Engineering Chemistry Research, vol. 43, no. 2, pp. 614–622, 2004.

10. J. Eßer, P. Wasserscheid, and A. Jess, "Deep desulfurization of oil refinery streams by extraction with ionic liquids," Green Chemistry, vol. 6, no. 7, pp. 316–322, 2004.

11. C. Huang, B. Chen, J. Zhang, Z. Liu, and Y. Li, "Desulfurization of gasoline by extraction with new ionic liquids," Energy and Fuels, vol. 18, no. 6, pp. 1862–1864, 2004.

12. Y. Nie, C. Li, A. Sun, H. Meng, and Z. Wang, "Extractive desulfurization of gasoline using imidazolium-based phosphoric ionic liquids," Energy and Fuels, vol. 20, no. 5, pp. 2083–2087, 2006.

13. I. Anugwom, P. Maki-Arvela, T. Salmi, and J. P. Mikkola, "Ionic liquid assisted extraction of nitrogen and sulfur-containing air pollutants from model oil and regeneration of the spent ionic liquid," Journal of Environmental Protection, vol. 2, no. 6, pp. 796–802, 2011.

14. W. Jian-long, D.-S. Zhao, E.-P. Zhou, and Z. Dong, "Desulfurization of gasoline by extraction with N-alkyl-pyridinium-based ionic liquids," Journal of Fuel Chemistry and Technology, vol. 35, no. 3, pp. 293–296, 2007.
15. A. Bösmann, L. Datsevich, A. Jess, A. Lauter, C. Schmitz, and P. Wasserscheid, "Deep desulfurization of diesel fuel by extraction with ionic liquids," Chemical Communications, no. 23, pp. 2494–2495, 2001.
16. G. W. Meindersma, A. Podt, and A. B. De Haan, "Selection of ionic liquids for the extraction of aromatic hydrocarbons from aromatic/aliphatic mixtures," Fuel Processing Technology, vol. 87, no. 1, pp. 59–70, 2005.
17. C. F. Poole and S. K. Poole, "Extraction of organic compounds with room temperature ionic liquids," Journal of Chromatography A, vol. 1217, no. 16, pp. 2268–2286, 2010.
18. M. J. Earle and K. R. Seddon, "Ionic liquids. Green solvents for the future," Pure and Applied Chemistry, vol. 72, no. 7, pp. 1391–1398, 2000.
19. R. Martínez-Palou and S. P. Flores, "Perspectives of ionic liquids applications for clean oilfield technologies," in Ionic Liquids: Theory, Properties, New Approaches, A. Kokorin, Ed., InTech, Rijeka, Croatia, 2011.
20. M. A. Cerón, D. J. Guzmán-Lucero, J. F. Palomeque, and R. Martínez-Palou, "Parallel microwave-assisted synthesis of ionic liquids and screening for denitrogenation of straight-run diesel feed by liquid-liquid extraction," Combinatorial Chemistry and High Throughput Screening, vol. 15, no. 5, pp. 427–432, 2012.
21. M. F. Casal, Desulfurization of fuel oils by solvent extraction with ionic liquids [Ph.D. thesis], University of Santiago de Compostela, Santiago de Compostela, Spain, 2010.
22. N. Papaiconomou, N. Yakelis, J. Salminen, R. Bergman, and J. M. Prausnitz, "Synthesis and properties of seven ionic liquids containing 1-methyl-3-octylimidazolium or 1-butyl-4-methylpyridinium cations," Journal of Chemical and

Engineering Data, vol. 51, no. 4, pp. 1389–1393, 2006.
23. P. Bonhôte, A. P. Dias, N. Papageorgiou, K. Kalyanasundaram, and M. Grätztel, "Hydrophobic, highly conductive ambient-temperature molten salts," Inorganic Chemistry, vol. 35, no. 5, pp. 1168–1178, 1996.
24. J. G. Huddleston, H. D. Willauer, R. P. Swatloski, A. E. Visser, and R. D. Rogers, "Room temperature ionic liquids as novel media for ‹clean› liquid-liquid extraction," Chemical Communications, no. 16, pp. 1765–1766, 1998.
25. B. M. Peric, E. Marti, J. Sierra, R. Cruanas, and M. A. Garau, "Green chemistry: ecotoxicity and biodegradability of ionic liquids," in Recent Advances in Pharmaceutical Sciences II, chapter 6, pp. 89–113, Transworld Research Network, Kerala, India, 2012.
26. A. Stanislaus, A. Marafi, and M. S. Rana, "Recent advances in the science and technology of ultra low sulfur diesel (ULSD) production," Catalysis Today, vol. 153, no. 1-2, pp. 1–68, 2010.

Chapter 3

Catalyst Deactivation and Engineering Control for Steam Reforming of Higher Hydrocarbons in a Novel Membrane Reformer

Zhongxiang Chen, Yibin Yan, and
Said S.E.H. Elnashaie

Department of Chemical Engineering, Auburn University, 230 Ross Hall, Auburn, AL 36849-5127, USA

ABSTRACT

The catalyst deactivation and reformer performance in a novel circulating fluidized bed membrane reformer (CFBMR) for steam reforming of higher hydrocarbons are investigated using mathematical models. A catalyst deactivation model is developed based on a random carbon deposition mechanism over nickel reforming catalyst. The results show that the reformer has a strong tendency for carbon formation and catalyst deactivation at low steam to carbon feed ratios (<1.4mol/mol) for high reaction temperatures (>700K) and high pressures (>506.5kPa). The trend is similar for the cases without and with hydrogen selective membranes. Based on this preliminary investigation, an engineering control approach, i.e., in-site control with a concept of critical/minimum steam to carbon feed ratio, is proposed and used to determine the carbon deposition free regions for both cases without and with hydrogen membranes. The comparison between the reported data and model simulation shows that the critical steam to carbon feed ratio predicted by the model agrees well with the reported industrial/experimental operating data.

INTRODUCTION

In recent years, considerable attention has been paid to the possibilities of utilizing the clean fuel hydrogen as an important energy source for the 21st century (Armor, 1999; Goltsov and Veziroglu, 2002; Ohi, 2002). In January 2003, President Bush announced in his State of the Union address $1.2 billion in research funding for developing clean, hydrogen-powered automobiles and fuel cells. Four major catalytic and non-catalytic approaches are widely used for hydrogen production, they are (1) steam reforming of hydrocarbons, (2) partial oxidation of heavy oil, (3) partial oxidation of coal and, (4) electrolysis of water (Scholz, 1993). Currently, steam reforming of hydrocarbons contributes about 50% of the world's hydrogen production (Scholz, 1993; Armor, 1999). The main advantages of the steam reforming process are: (1) it extracts the hydrogen not only from the hydrocarbons but also from the water (water resource is inexhaustible) and; (2) the reaction rate is very fast, although limited by thermodynamic equilibrium. However, this process is accompanied by unfavorable and undesired formation of different carbonaceous deposits or coke, which deactivates the catalyst, it can even destroy the reformer

(Rostrup-Nielsen 1979 and Rostrup-Nielsen 1997; Borowiecki et al., 1997; Bartholomew, 2001; Olsbye et al., 2002). By burning-off the deposited carbon with air or oxygen the catalyst is regenerated. The catalyst may be permanently deactivated by sintering (loss of surface area) during the catalyst regeneration, therefore careful control of burn-off is necessary (Trimm, 1984). The carbon formation and catalyst deactivation during steam reforming of hydrocarbons have been intensively studied and different approaches have been developed for controlling the carbon formation. For example, it is possible to use potassium, magnesia, urania or molybdenum to improve the reforming catalysts by inhibiting the carbon formation or promoting the carbon gasification (Borowiecki et al., 1997; Trimm, 1999; Kepinski et al., 2000). Carbon formation can also be avoided by using high steam to carbon (of hydrocarbon) feed ratios (Twigg, 1989; Elnashaie and Elshishini, 1993; Christensen, 1996; Bartholomew, 2001). However, there is still no generally accepted model to describe the carbon formation and catalyst deactivation due to the complexity of the reforming process (Ren et al., 2002). Experimental data and theoretical analysis have shown that the carbon formation rate is largely dependent on the catalyst chemical composition as well as its preparation procedure (Rostrup-Nielsen, 1974; Borowiecki, 1987; Forzatti and Lietti, 1999). In 1945 Voorhies empirically described the carbon formation as a function of reaction time by the following equation (Vooehies, 1945):

$$C = k_c t^n, \tag{1}$$

Where C is the concentration of carbon formed on the catalyst, t is the reaction time, k_c is the carbon formation rate constant and n is an exponent (usually <1). Rostrup-Nielsen (1974) studied the carbon formation from higher hydrocarbons in a thermogravimetric system and suggested that the amount of carbon formed may be empirically expressed by the following equation:

$$C = k_c(t - t_0), \tag{2}$$

Where, t_0 is the induction time for carbon formation on the catalyst. In these equations, the amount of carbon formed on the catalyst is assumed to be independent of the partial pressure of hydrocarbons, which is a very unrealistic assumption. In this paper a random carbon deposition mechanism is suggested for nickel reforming catalyst and then a catalyst deactivation model is developed, which is incorporated into a set of model equations to study the catalyst deactivation and engineering control for steam reforming of higher hydrocarbons in an earlier proposed circulating fluidized bed membrane reformer (CFBMR) (Chen and Elnashaie, 2002; Chen 2002 and Chen 2003a). Fig. 1 shows a complete schematic diagram of the proposed novel process. Inside the CFBMR there are a number of ceramic membrane tubes coated by thin palladium layer and/or dense perovskite oxygen selective membrane tubes. Between these membrane tubes the nickel reforming catalyst is fast fluidized and steam reforming of higher hydrocarbons takes place. The product hydrogen permeates selectively through hydrogen membranes and then it is carried away by sweep gas such as steam in the hydrogen membrane tubes. Air is fed into the oxygen selective membrane tubes where oxygen permeates into the reaction side for oxidative reforming of hydrocarbons. The deactivated catalyst is carried out of the reformer with the exit gases, regenerated in a catalyst regenerator by burning off the deposited carbon using excess air. Then the regenerated catalyst is separated from the gas stream in a gas–solid separator and finally recycled to the riser reformer. Because the effluent gases from the gas–solid separator are rich in carbon dioxide (CO_2), the main greenhouse gas causing global warning. In order to avoid/control this polluting emission and reduce the negative effect on the environment, as shown in the right of Fig. 1, a promising approach of dry reforming of methane is used in the downstream to capture the CO_2 for syngas production, which can be further converted into valuable fuel additives or chemicals such as methanol (El Solh et al., 2001; Verykios, 2003). In this paper only the left part CFBMR is investigated for the cases without and with hydrogen selective membranes.

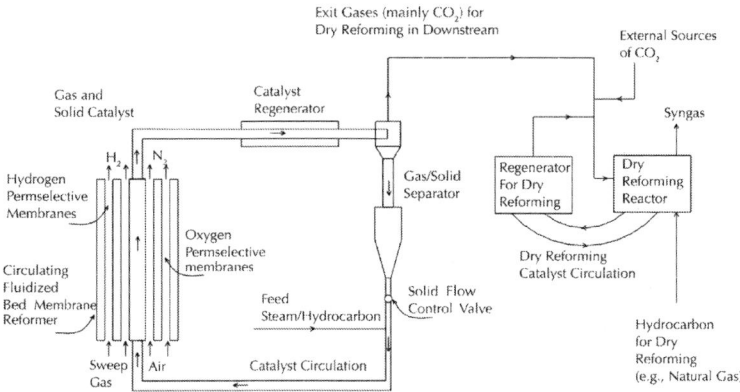

Figure 1: Schematic diagram of the novel process containing a CFBMR.

RANDOM CARBON DEPOSITION AND CATALYST DEACTIVATION MODEL

During the steam reforming of hydrocarbons on nickel reforming catalyst, three typical kinds of carbon species were identified: pyrolytic carbon, encapsulating carbon and whisker or filamentous carbon (Rostrup-Nielsen, 1979; Forzatti and Lietti, 1999; Trimm, 1999; Bartholomew, 2001). Pyrolytic carbon is usually obtained by thermal cracking of hydrocarbons above 600°C and deposition of carbon precursors. Encapsulating carbon is formed by slow polymerization of unsaturated hydrocarbons below 500°C. Whisker carbon is produced by diffusion of carbon into nickel crystals, detachment of nickel from the support and growth of whiskers with nickel on the top of the catalyst above 450°C. Both pyrolytic and encapsulating carbons cover the catalyst particle surface and therefore deactivate the catalyst. Although whisker carbon does not deactivate the catalyst directly, the accumulation of whisker carbon blocks the catalyst pores and increases the pressure drop to unacceptable levels in the reformers. Trimm (1984) suggested that the production of catalytic carbon on the nickel catalyst could best be described with the aid of Fig. 2: hydrocarbons adsorb on the catalyst surface and may react to produce gas phase products or dehydrogenated intermediates. This process continues until carbon is produced on the surface, which in turn can isomerize to other forms of carbon.

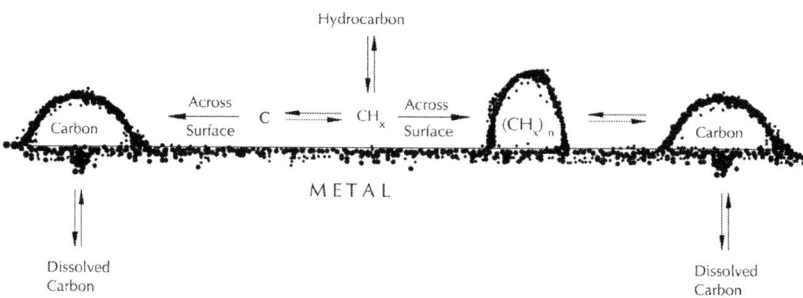

Figure 2: Mechanism of carbon formation during steam reforming of hydrocarbons (from Trimm, 1984).

Based on the above mechanism of different carbon formation over nickel reforming catalyst, a random carbon deposition and catalyst deactivation model is proposed as follows: Hydrocarbon adsorbs on the nickel reforming catalyst surface and may react to produce gas phase products or to form precursors of carbon or protocoke, which is mobile and then deposits randomly on the nickel catalyst, either on the coked or uncoked sites. The coking process can be schematically presented as follows:

$$C_nH_m \xrightarrow{r} CH_x(\text{protocoke}) \xrightarrow{r_d} \text{coke}. \tag{3}$$

Let C, C^* be the concentrations of the deposited carbon and protocoke on the catalyst; r, r_d be the protocoke formation rate and carbon deposition rate; S, S_0 be the concentrations of active sites per gram of catalyst at reaction time t and initially, respectively. Starting with the material balance for protocoke and deposited carbon on the catalyst, we get the following two equations:

$$\frac{dC^*}{dt} = r(S, P_{C_nH_m}, \ldots) - r_d(S_0, C^*), \tag{4}$$

$$\frac{dC}{dt} = r_d(S_0, C^*). \tag{5}$$

In Eq. (4) the protocoke formation rate r involves the number of active sites at time t because the protocoke is formed from hydrocarbon, which is based on the available active sites. While in Eq. (5) the carbon deposition rate r_d involves the initial active sites S_0 and not S because of the assumption that protocoke can deposit either on the coked or uncoked sites. Considering the growth of protocoke into coke is fast and using the approximation of pseudo-steady-state for protocoke, we get

$$\frac{dC^*}{dt} = r(S, P_{C_nH_m}, \ldots) - r_d(S_0, C^*) = 0. \tag{6}$$

Since protocoke is formed from the adsorbed hydrocarbon on the catalyst, the formation rate could be presumed proportional to the concentration of adsorption sites. Then Eq. (5) may be rewritten as

$$\frac{dC}{dt} = r(S, P_{C_nH_m}, \ldots) = \frac{S}{S_0} r_0(S_0, P_{C_nH_m}, \ldots), \tag{7}$$

Where r_0 is the initial protocoke formation rate. This equation proposes that the coke deposition rate in general depends on the concentration of the active sites S and S_0. If the active site coverage is the main cause of the deactivation of catalyst for steam reforming of hydrocarbons, the catalyst activity function φ (sometimes it is also called the catalyst deactivation function) may be defined by

$$\phi = S/S_0. \tag{8}$$

Because the change of active sites on the catalyst equals to the loss of active sites caused by the carbon deposition, we can get

$$W_{cat} \Delta S = [-r_d(S_0, C^*)] W_{cat} \Delta t \alpha \frac{S}{S_0}, \tag{9}$$

Where W_{cat} is the catalyst weight in grams; Δt is the segment of reaction time in seconds; α is a conversion coefficient from coke concentration C to active site concentration S.

When Δt 0, we get the following differential equation:

$$\frac{dS}{dt} = \alpha \frac{S}{S_0}[-r_d(S_0, C^*)]. \tag{10}$$

Substituting Eq. (5) into Eq. (10), we get

$$\frac{dS}{dt} = -\alpha \frac{S}{S_0}\frac{dC}{dt}. \tag{11}$$

Finally, the catalyst deactivation caused by the carbon deposition may be represented by the following equation after integrating Eq. (11):

$$\phi = \exp(-\alpha_C C), \tag{12}$$

Where $\alpha_C \equiv \alpha/S_0$ is the catalyst deactivation constant. Assume that when the catalyst pores are fully filled with carbon, the catalyst will deactivate completely. For nickel catalysts, typical catalyst surface area is in the range of 20–66m²/g-catalyst (Tottrup, 1982; Xu and Froment, 1989; Sehested et al., 2001) and pore mean radius is around 17Å (Biswas and Do, 1987). We assume the pores are equivalent as cylinders. In the proposed CFBMR, fine catalyst particles (186μm) are used for free circulation. The reported density of industrial nickel catalyst is 2835kg/m³ (Elnashaie and Elshishini, 1993) and the density of coke is 440.5kg/m³ (Perry et al., 1984). Then the maximum amount of carbon that can be deposited on the catalyst is 0.16g/g-catalyst. Ren et al. (2002) reported the experimental maximum coke content to be 16wt% for naphtha reforming catalyst Pt–Re/Al_2O_3. Forzatti and Lietti (1999) reported that the coke deposition on the reforming catalyst may amount to 15–20% (w/w) of the catalyst. Although the specific compositions of reforming catalyst are different, the maximum carbon content estimated above is quite close to the reported data. For mathematical simplicity, suppose the catalyst will lose its 99% activity (or is 0.01) when the carbon content reaches 0.16g/g-catalyst. Then the catalyst deactivation constant $_c$ is 28.8g-catalyst/g-carbon. Therefore without carbon deposition, the catalyst does not deactivate and the catalyst activity is 1.0. When carbon deposition increases, the catalyst deactivates and the catalyst activity decreases. The larger the carbon deposition on the catalyst, the lower the catalyst activity function and the more significant the catalyst deactivation.

REACTIONS AND KINETICS

Many researchers used heptane as a model component for steam reforming of higher hydrocarbons (Rostrup-Nielsen, 1974; Tottrup, 1982; Christensen, 1996). In this investigation we also use heptane as a model component for higher hydrocarbons. The possible reactions and their kinetics are summarized in Table 1, which are carefully chosen from the open literatures except for the rate equation of heptane cracking. As mentioned earlier, Rostrup-Nielsen (1974) suggested a kinetic rate equation (Eq. (2)) that the amount of carbon formed on the catalyst is assumed to be independent of the partial pressure of hydrocarbons. Taking into consideration of the effect of heptane feed, we empirically correlated the Rostrup-Nielsen's experimental data and obtained the carbon formation rate equation from heptane, which is marked with a "*" in Table 1. All the reaction rates r_1 to r_9 are for reactants, as the reaction expression shown in Table 1. The carbon formation by the decomposition of carbon monoxide, i.e., the Boudouard reaction, is usually regarded as a reversible reaction. However, during the steam reforming of higher hydrocarbons, the carbon formation from hydrocarbon and methane are more important than that from carbon monoxide. Furthermore, at the presence of steam and hydrogen, carbon gasification by steam and hydrocarbon are also more important than that by CO_2. On the other hand, the kinetics of Boudouard reaction for carbon formation on the nickel catalyst reported by Tottrup (1976) is an irreversible rate equation. In order to address this, we use another reaction rate equation r_9 shown in Table 1 for the reverse of the Boudouard reaction.

Table 1: Reactions and kinetic rate equations

Reaction	Kinetic equation	Reference
$C_7H_{16} + 7H_2O \rightarrow 7CO + 15H_2$	$r_1 = \dfrac{k_1 P_{C_7H_{16}}}{\left[1 + 25.2 \dfrac{P_{C_7H_{16}} P_{H_2}}{P_{H_2O}} + 0.077 \dfrac{P_{H_2O}}{P_{H_2}}\right]^2}$	Tottrup (1982)
$CO + 3H_2 \rightleftharpoons CH_4 + H_2O$	$r_2 = k_2 \left(\dfrac{P_{CH_4} P_{H_2O}}{P_{H_2}^{2.5}} - \dfrac{P_{CO} P_{H_2}^{0.5}}{K_2} \right) / DEN^2$	Xu and Froment (1989)
$CO + H_2O \rightleftharpoons CO_2 + H_2$	$r_3 = k_3 \left(\dfrac{P_{CO} P_{H_2O}}{P_{H_2}} - \dfrac{P_{CO_2}}{K_3} \right) / DEN^2$	Xu and Froment (1989)
$CH_4 + 2H_2O \rightleftharpoons CO_2 + 4H_2$	$r_4 = k_4 \left(\dfrac{P_{CH_4} P_{H_2O}^2}{P_{H_2}^{3.5}} - \dfrac{P_{CO_2} P_{H_2}^{0.5}}{K_2 K_3} \right) / DEN^2$	Xu and Froment (1989)
$C_7H_{16} \rightarrow 7C + 8H_2$	$r_5 = k_5 P_{C_7H_{16}}^{0.569\,a}$	Rostrup-Nielsen (1974)
$CH_4 \rightleftharpoons C + 2H_2$	$r_6 = \dfrac{k_6 K_{CH_4} \left(P_{CH_4} - \dfrac{P_{H_2}^2}{K_{6a}} \right)}{\left(1 + \dfrac{P_{H_2}^{1.5}}{K_{6b}} + K_{CH_4} P_{CH_4} \right)^2}$	Snoeck et al. (1997)
$2CO \rightarrow C + CO_2$	$r_7 = \dfrac{k_7 P_{CO}}{\left(1 + K_{7a} P_{CO} + K_{7b} \dfrac{P_{CO_2}}{P_{CO}} \right)^2}$	Tottrup (1976)
$C + H_2O \rightarrow CO + H_2$	$r_8 = k_8 P_{H_2O}^{0.5}$	Chen et al. (2000)
$C + CO_2 \rightarrow 2CO$	$r_9 = k_9 P_{CO_2}^{0.5}$	Chen et al. (2000)
where, $DEN = 1 + K_{CO} P_{CO} + K_{H_2} P_{H_2} + K_{CH_4} P_{CH_4} + K_{H_2O} P_{H_2O}/P_{H_2}$		

a Empirically obtained from the experimental data reported by Rostrup-Nielsen (1974).

MATHEMATICAL MODELING AND SIMULATION CONDITIONS

Due to the high gas–solid velocity (~3m/s) in the circulating fluidized bed reformer, we assume that plug flow model applies in this novel CFBMR for both the gas and solid phases. Thus the CFBMR is modeled as a plug flow reactor (PFR) with co-current flow in the reactor and membrane sides. The other major assumptions for the mathematical model are as follows:

- Steady-state operation in the reaction and hydrogen membrane sides.
- The palladium based composite membranes are 100% selective for hydrogen only.
- There is no slip between the solid and gases, both are in plug flow.
- The heat capacities of the components are constant.
- The reformer and hydrogen membranes are operated at constant pressure.
- The reformer is simulated under isothermal conditions.
- The deactivated catalyst is fully regenerated before recycling to the riser reformer.
- The hydrogen selective membranes are not affected by carbon formation.

The steady-state model equations for material balance in reaction side are given by

$$\frac{dF_i}{dl} = \rho_C(1-\varepsilon)A_f \sum_{j=1}^{9} \sigma_{i,j} r_j - aJ_{H_2}\pi N_{H_2} d_{H_2}, \tag{13}$$

Where a is a control index, when component i is hydrogen, $a=1$, otherwise, $a=0$.

The material balance equation in hydrogen selective membrane tubes is given by

$$\frac{dF_{H_2,P}}{dl} = \pi N_{H_2} d_{H_2} J_{H_2}. \tag{14}$$

For palladium based hydrogen selective membranes, the hydrogen permeation flux can be calculated by the following equation (Shu and Kaliaguine, 1994; Barbieri and Di Maio, 1997):

$$J_{H_2} = \frac{2.003 \times 10^{-5}}{\delta_{H_2}} \exp\left(\frac{-15,700}{RT}\right)$$

$$\times \left(\sqrt{p_{H_2,r}} - \sqrt{p_{H_2,p}}\right) \frac{\text{mol}}{\text{m}^2 \text{ s}}. \tag{15}$$

Since the catalyst deactivation occurs when carbon deposits, the rate equations are reformulated accordingly by introducing the catalyst activity function to the reaction rates as follows:

$$r_j = r_{j0} \phi_j, \tag{16}$$

Where ϕ_j is the specific catalyst activity function for the jth reaction, which is calculated using Eq. (12) or equal to 1.0 depending on whether the jth reaction is affected by the catalyst deactivation. In this preliminary investigation, only the last two reaction rate equations r_8 and r_9 are considered unaffected by the catalyst deactivation because the uncatalyzed carbon gasification kinetics are used in Table 1. r_{j0} is the initial reaction rate with fresh catalyst.

Using the above reaction kinetic equations, catalyst activity function and reactor model equations, the catalyst deactivation and hydrogen production in the CFBMR are investigated. Unless otherwise specified, the simulation is performed at the following standard conditions summarized in Table 2.

Table 2: Standard simulation conditions

CFBMR construction parameters and catalyst properties	
Length of the reformer and membrane tubes	2m
Diameter of the reformer	0.0978m[a]
Diameter of hydrogen selective membrane tubes	0.00498m[b]
Thickness of palladium layer on hydrogen membrane tubes	20μm[c]
Catalyst particle density	2835kg/m³[a]
Mean catalyst particle size	186μm[b]
Solid fraction in circulating fluidization bed	0.2[d]

Process gas feed and reaction conditions [e]	
Heptane feed rate	0.178mol/s
Steam feed rate	2.5mol/s
Steam to carbon ratio	2mol/mol
Reaction temperature	823K
Reaction pressure	1013kPa
Pressure in hydrogen selective membrane tubes	101.3kPa
Sweep gas feed rate in hydrogen selective membrane tubes	0.278mol/s

[a]Based on Elnashaie and Elshishini (1993).

[b]Based on Adris et al. (1994).

[c]Based on Shu and Kaliaguine (1994).

[d]Based on Kunii and Levenspiel 1990 and Kunii and Levenspiel 1997.

[e]Based on Tottrup (1982) and checked the CFBMR is simulated at circulating fluidization regime.

RESULTS AND DISCUSSION

Catalyst Deactivation and CFBMR Performance without Hydrogen Selective Membranes

In this paper the catalyst deactivation and CFBMR performance for the steam reforming of heptane is investigated. The simulation is performed under isothermal condition. As shown later some scales of the three-dimension plots are reversed in the direction for the better view purpose. First, the catalyst deactivation and CFBMR performance are investigated at 1013kPa with different steam to carbon (S/C) feed ratios and reaction temperatures for the case without hydrogen selective membranes. Fig. 3 shows the catalyst activity as a function of steam to carbon (of heptane) feed ratio and reaction temperature under the reaction pressure of 1013kPa. The investigated range of S/C feed ratio is 0–4.4mol/mol and the range of temperature is 623–923K. Fig. 4 shows the carbon content on the catalyst for this investigation. At low S/C feed ratios and high reaction temperatures, the nickel reforming catalyst is deactivated significantly since a lot of heptane cracks to form carbon. The catalyst activity can be as low as 0.69 shown in Fig. 3. The carbon content on the catalyst shown in Fig. 4 is up to 0.0130g/g-catalyst or 1.3wt% for the case without hydrogen selective membranes. At lower temperatures 623–683K, the catalyst deactivation is negligible because of the limited carbon deposition at these lower temperatures, as shown in Fig. 4. The catalyst activity increases with the increase of S/C feed ratio and with the decrease of reaction temperature. At the corner where S/C feed ratio is less than 1.4mol/mol and the reaction temperature is higher than 700K, the reforming reactions have a strong tendency for carbon deposition on the nickel reforming catalyst and therefore causing significant catalyst deactivation shown in Fig. 3.

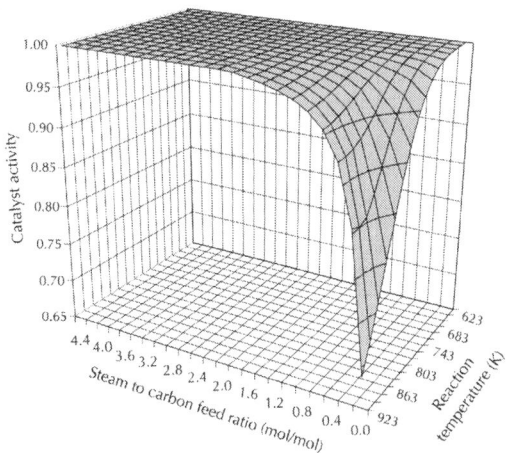

Figure 3: Catalyst activity as a function of steam to carbon feed ratio and reaction temperature for the case without hydrogen selective membranes at 1013kPa.

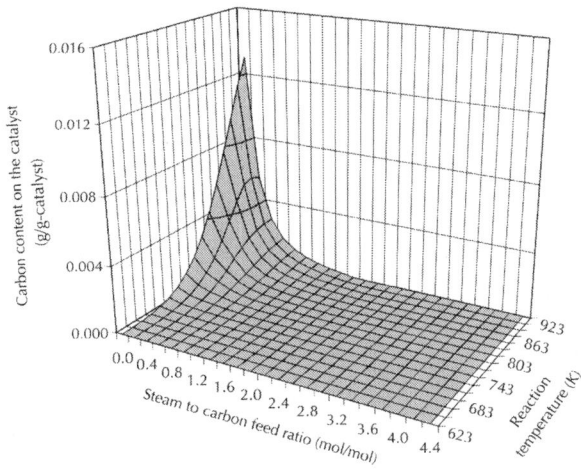

Figure 4: Carbon content on the catalyst as a function of steam to carbon feed ratio and reaction temperature for the case without hydrogen selective membranes at 1013 kPa.

Fig. 5 and Fig. 6 show the conversion of heptane (defined as the total moles of heptane converted per mol of heptane fed) and total yield of hydrogen (defined as the total moles of hydrogen produced, both in reaction side and hydrogen selective membrane tubes, per mole of heptane fed) under different S/C feed ratios and reaction temperatures. For most reaction conditions heptane is fully converted by steam reforming and heptane cracking. At lower temperatures and/or lower S/C feed ratios, for example, at 623–723K and/or at S/C feed ratio of 0–1.0 mol/mol, the conversion of heptane is relatively small. The conversion of heptane increases when reaction temperature or S/C feed ratio increases. Fig. 6 shows that the total yield of hydrogen is non-monotonic with respect to the reaction temperature when the S/C feed ratio is higher than 0.4 mol/mol. This phenomenon has been extensively investigated by Chen et al. (2003b). It is mainly caused by the strong methanation reaction around 723K in the steam reforming of heptane system. At low temperatures such as 623–723K, the steam reforming of heptane reaction dominates the system and the methanation reaction is relatively negligible. Around 723K the methanation reaction becomes more significant and consumes a lot of hydrogen produced from heptane steam reforming, causing a large decrease in the yield of hydrogen. However, at high temperatures such as 823K or higher, the steam reforming of methane, the reverse process of methanation, become more and more important, thus it decreases the formation of methane and enhances the production of hydrogen. As a result, the non-monotonic behavior in the yield of hydrogen with respect to the reaction temperature appears (Chen et al., 2003b). Fig. 7 shows the yield of methane in the reforming system. The higher the tendency for methanation, the lower the yield of hydrogen. For heptane steam reforming, the maximum theoretical yield of hydrogen is 22 according to the following complete reforming reaction in which the final products are CO_2 and hydrogen:

$$C_7H_{16} + 14H_2O \rightarrow 7CO_2 + 22H_2. \quad (17)$$

Obviously, because of the reversibility associated with the methane steam reforming reaction (or methanation) and water gas shift reaction, the production of hydrogen is usually limited by thermodynamic equilibrium, resulting in low yield of hydrogen, even at high S/C feed ratio. For example, at S/C feed ratio of 4.0mol/mol, the total yields of hydrogen are 13.624 at 623K, 2.844 at 723K, 6.650 at 823K and 11.760 at 923K, respectively.

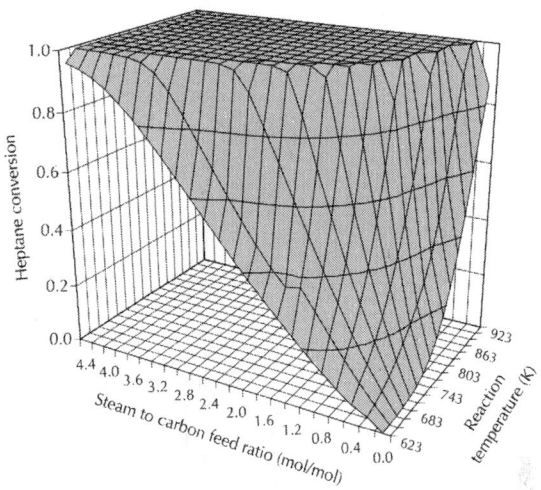

Figure 5: Heptane conversion as a function of steam to carbon feed ratio and reaction temperature for the case without hydrogen selective membranes at 1013 kPa.

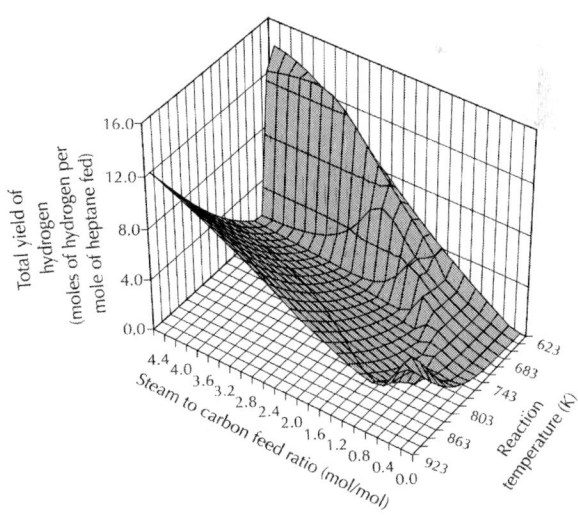

Figure 6: Total yield of hydrogen as a function of steam to carbon feed ratio and reaction temperature for the case without hydrogen selective membranes at 1013 kPa.

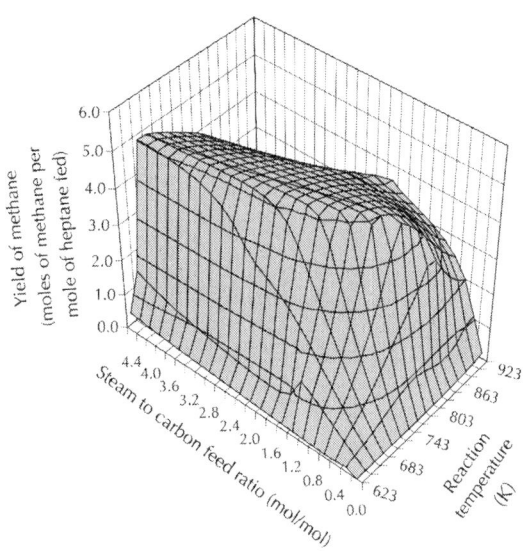

Figure 7: Yield of methane as a function of steam to carbon feed ratio and reaction temperature for the case without hydrogen selective membranes at 1013 kPa.

Next, the catalyst deactivation and CFBMR performance are investigated at 823K with different S/C feed ratios and reaction pressures for the case without hydrogen selective membranes. The investigated range of S/C feed ratio is the same, 0–4.4mol/mol. The range of reaction pressure is 101.3–3140.3kPa. Fig. 8 shows the catalyst activity as a function of S/C feed ratio and reaction pressure at 823K. For most of the operation conditions, the catalyst activity is high (close to 1.0), which means low carbon content on the reforming catalyst and insignificant catalyst deactivation. At S/C feed ratio of 1.4mol/mol or higher, the catalyst activity is higher than 0.972. Accordingly, the carbon content on the catalyst is below 0.001g/g-catalyst or 0.1wt%%. At a corner where S/C feed ratio is less than 1.4mol/mol and reaction pressure is higher than 506.5kPa, the reforming reactions have a strong tendency for carbon formation and deposition on the nickel reforming catalyst, causing a significant catalyst deactivation. The carbon content can reach as high as 0.0181g/g-catalyst at 3140.3kPa for the special condition without steam (i.e., S/C feed ratio is 0mol/mol), resulting in a very low catalyst activity, about 0.594. The catalyst activity increases when the S/C feed ratio increases or when the reaction pressure decreases. At high S/C feed ratio, the steam reforming of hydrocarbons (heptane and by-product methane) dominates the reforming system and the high excess steam feed enhances the carbon gasification, which suppresses the carbon deposition on the nickel reforming catalyst and therefore decreases the catalyst deactivation.

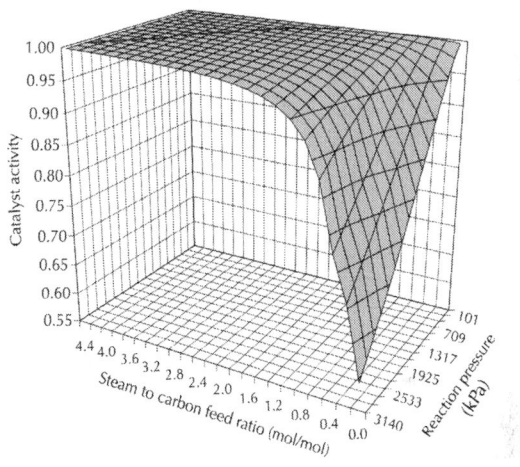

Figure 8: Catalyst activity as a function of steam to carbon feed ratio and reaction pressure for the case without hydrogen selective membranes at 823K.

Fig. 9 shows the total yield of hydrogen as a function of S/C feed ratio and reaction pressure at 823K for the case without hydrogen membranes. At very low S/C feed ratio, for example, 0–0.4mol/mol, the yield of hydrogen increases when the reaction pressure increases, while above 0.4mol/mol, the yield of hydrogen increases with the increase of the S/C feed ratio and decreases with the increase of the reaction pressure. This can be explained as follows: at low S/C feed ratio 0–0.4mol/mol, the heptane cracking reaction is dominating in the system. Because this cracking reaction is irreversible, the high pressure will not limit the conversion of heptane for cracking. Since the reaction order of the cracking of heptane with respect to the partial pressure of heptane is 0.569 shown in Table 1, high operating pressure gives high cracking rate of heptane. As a result, the yield of hydrogen increases when the reaction pressure increases. However, when S/C feed ratio increases, the steam reforming of heptane becomes important and also the methanation and water gas shift reactions. The methanation or steam reforming of methane and water gas shift reaction are fast reversible reactions, which are strongly affected by the thermodynamic equilibrium. Since steam reforming of heptane is accompanied with an increase in molecule number, the higher the operating pressure, the higher the effect of the thermodynamic equilibrium. Thus the yield of hydrogen decreases when the reaction pressure increases. This phenomenon also indicates that the thermodynamic equilibrium limits the production of hydrogen in steam reforming system.

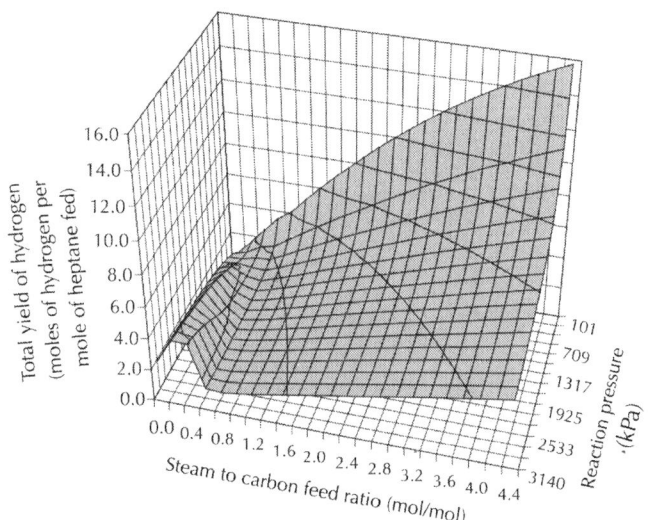

Figure 9: Yield of hydrogen as a function of steam to carbon feed ratio and reaction pressure for the case without hydrogen selective membranes at 823K.

Catalyst Deactivation and CFBMR Performance with Hydrogen Selective Membranes

In this section 20 hydrogen selective membranes are used to investigate their effect on the catalyst deactivation and CFBMR performance. Fig. 10 shows the catalyst activity for this case. The trend is similar to the earlier case without hydrogen selective membranes. The catalyst is also deactivated significantly at low S/C feed ratio and high reaction temperatures. At most reforming conditions the effect of carbon deposition on the catalyst deactivation is negligible. While at the corner where S/C feed ratio is less than 1.6mol/mol and the reaction temperature is higher than 700K, the catalyst activity decreases significantly with the decrease of the S/C feed ratio and with the increase of the reaction temperature. The difference between both cases without and with hydrogen selective membranes is only the magnitude of the catalyst activity. With 20 hydrogen membranes the lowest catalyst activity is 0.59. Accordingly, the maximum carbon content on the catalyst is 0.0183g/g-catalyst or 1.8wt%. While in the earlier case without hydrogen selective membranes the lowest catalyst activity function is 0.69 shown in Fig. 3 or the maximum carbon content is 0.0130g/g-catalyst (1.3wt %) shown in Fig. 4. The carbon content for the case with 20 hydrogen selective membranes increases by 40.8%. The result can be explained by the effect of hydrogen selective membranes on the reactions, especially on the carbon formation and carbon gasification in the CFBMR. Table 3 summarized the possible effects of hydrogen selective membranes on the directions of different reactions. For those irreversible reactions, the reaction direction will not be affected by the use of hydrogen selective membranes. While for the reversible reactions, the use of hydrogen membranes will be favorable or "shift" some reactions to certain directions as clearly shown in Table 3. Although the use of the hydrogen selective membranes will be favorable for the steam reforming of methane (by-product of steam reforming of heptane via methanation) or suppress the formation of methane, the strong methanation can still produce much methane for carbon formation, leading to a little higher carbon content on the nickel reforming catalyst. At the extreme condition without steam, i.e., S/C feed ratio of 0mol/mol, the carbon produced by heptane cracking will not be gasified by steam. However, the product hydrogen can react with carbon to form methane at high temperatures. The use of hydrogen selective membranes decreases the concentration of hydrogen in the system, as a result the carbon gasification by hydrogen is lower and the carbon content is higher for the case with hydrogen selective membranes. Therefore, the catalyst activity is lower than the case without hydrogen selective membranes, which is shown in Fig. 10. As mentioned above, the use of hydrogen selective membranes "shifts" the reversible reactions to the direction for hydrogen production. Therefore, the yield of hydrogen shown in Fig. 11 is significantly improved using hydrogen selective membranes. For example, at S/C feed ratio of 2mol/mol with 823K and 1013kPa, the yield of hydrogen is 3.726 for the case without hydrogen selective membranes and 19.298 for the case with hydrogen selective membranes, which is close to the theoretical maximum yield of hydrogen 22 shown by Eq. (17). The improvement is about 418% due to the "break" of the thermodynamic equilibrium limitation.

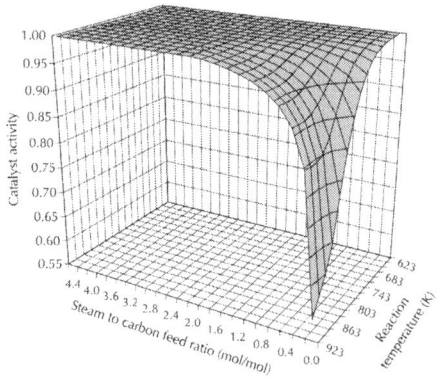

Figure 10: Catalyst activity as a function of steam to carbon feed ratio and reaction temperature for the case with 20 hydrogen selective membranes at 1013kPa.

Table 3: Effect of hydrogen selective membranes on the reaction directions

Reactions	Effect of hydrogen membranes on the reaction direction
$C_7H_{16}+7H_2O \rightarrow 7CO+15H_2$	NA[a]
$CO+3H_2 \rightleftharpoons CH_4+H_2O$	←[b]
$CO+H_2O \rightleftharpoons CO_2+H_2$	→
$CH_4+2H_2O \rightleftharpoons CO_2+4H_2$	→
$C_7H_{16} \rightarrow 7C+8H_2$	NA
$CH_4 \rightleftharpoons C+2H_2$	→
$2CO \rightarrow C+CO_2$	NA
$C+H_2O \rightarrow CO+H_2$	NA
$C+CO_2 \rightarrow 2CO$	NA

[a] NA=not affected.

[b] Left or right arrows mean "favorable" for this reaction direction.

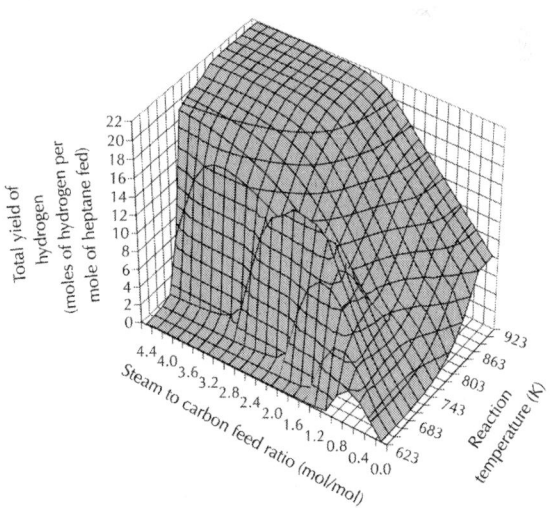

Figure 11: Yield of hydrogen as a function of steam to carbon feed ratio and reaction temperature for the case with 20 hydrogen selective membranes at 1013kPa.

Fig. 12 shows the catalyst activity at 823K as the function of S/C feed ratio and reaction pressure, respectively. The catalyst activity profile is similar to that shown in Fig. 8. While the lowest catalyst activity is 0.467 at S/C feed ratio of 0mol/mol and 3140kPa. The carbon content at this condition is 0.0264g/g-catalyst for the case with hydrogen selective membranes. The carbon content is 45.9% higher than the case without hydrogen selective membranes at the same operation condition. Fig. 13shows the total yield of hydrogen as a function of S/C feed ratio and reaction pressure for the case with hydrogen selective membranes. Compared to the hydrogen yield profile shown in Fig. 9, obviously, the thermodynamic equilibrium limitation for hydrogen production due to the high operating pressure is eliminated with hydrogen selective membranes. For example, at S/C feed ratio of 2mol/mol and 3140kPa, the total yield of hydrogen is 2.317 for the case without hydrogen selective membranes, while with hydrogen selective membranes, the total yield of hydrogen is 20.687. In Fig. 9 the yield of hydrogen decreases when operating pressure increases due to the fact that steam reforming of heptane is accompanied with an increase in molecule number. However, using hydrogen selective membranes, the yield of hydrogen increases when operating pressure increases. The improvement using hydrogen selective membranes at high reaction pressure is significant. Fig. 13 also shows that the yield of hydrogen increases when the S/C feed ratio increases. However, at high S/C feed ratios, for example, 3.0mol/mol or higher, the difference in hydrogen yield with different operating pressures is rather small. The total yield of hydrogen shown in Fig. 13 is very close to the theoretical maximum yield of hydrogen of 22 at the region where S/C feed ratio is higher than 3 mol/mol and reaction pressure is higher than 50605kPa.

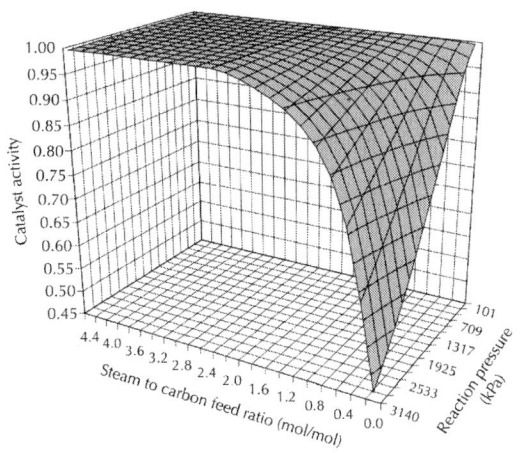

Figure 12: Catalyst activity as a function of steam to carbon feed ratio and reaction pressure for the case with 20 hydrogen selective membranes at 823K.

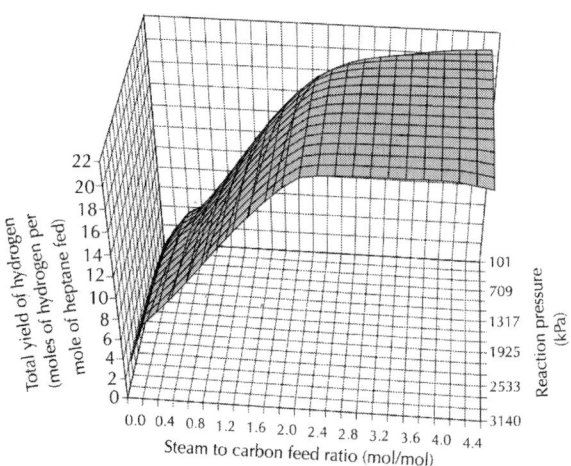

Figure 13: Total yield of hydrogen as a function of steam to carbon feed ratio and reaction pressure for the case with 20 hydrogen selective membranes at 823K.

Engineering Control for Carbon Deposition and Catalyst Deactivation

As mentioned earlier, there are several approaches to control the carbon formation on the steam reforming catalyst. These approaches may be classified into three groups according to their control stages. The first one is the well-designed optimal catalyst containing small amount of dopants such as Pt, Ir, Sn, Pb, Ge, As, Bi, Mo, Ag, etc. (Trimm, 1999). We may call it pre-reforming control or catalyst control. It usually takes a relatively long time to obtain such optimal catalysts. The second one is the in-site carbon formation suppression or gasification by steam, hydrogen, oxygen, etc. We may call it in-site control.

The third one is the deactivated catalyst regeneration such as burn-off using oxygen or air in the regenerator. We may call it post-reforming control. All of them are widely used in the catalyst design, production, utilization and regeneration. In this section we focus on the second approach, the in-site control. As shown earlier, the carbon content on the catalyst can be well-controlled without significant catalyst deactivation at certain high S/C feed ratios. In order to provide a practical carbon free reforming condition for the novel CFBMR, we investigated the catalyst activity as a function of reaction temperature, pressure and S/C feed ratio. The carbon deposition free boundary is defined as the critical/minimum S/C feed ratio that makes the carbon content on the catalyst practically negligible (close to zero), which means that the catalyst activity is very high (close to 1.0). Through preliminary investigation, the critical catalyst activity for this kind of carbon deposition free boundary simulation is chosen 0.995. Then by incrementing the S/C feed ratio in small steps at a given reaction temperature and pressure (other reforming conditions are the same as listed inTable 2), the point at which the catalyst activity is equal to critical value of 0.995 can be precisely determined. Thus it is possible to describe the carbon deposition free boundary for heptane steam reforming in CFBMR by determining a series of critical S/C feed ratios as functions of reaction temperature and reaction pressure.

Fig. 14 and Fig. 15 show the critical S/C feed ratios for the cases without and with 20 hydrogen selective membranes. In these two figures, the region above the surface can be considered as the carbon deposition free zone. While below this surface it is considered as the carbon deposition zone in which the catalyst activity is smaller than 0.995 and the catalyst deactivates significantly. Thus it is possible to use these findings to guide the practical operations for the novel CFBMR, especially with regard to the carbon formation and catalyst deactivation. Generally, the critical S/C feed ratio increases with the increases of the reaction temperature and reaction pressure. Fig. 16 shows the difference of critical S/C feed ratios between these two cases with and without 20 hydrogen selective membranes. At lower reaction temperatures 623–723K and lower pressures 101–1317kPa, the critical S/C feed ratios for the case with hydrogen selective membranes are higher than the case without hydrogen selective membranes. While at the other conditions where reaction temperature is higher than 723K and pressure is higher than 1317kPa, the critical S/C feed ratios for the case with hydrogen selective membranes are smaller than the case without hydrogen selective membranes. This interesting phenomenon can be explained by the effect of the removal of product hydrogen on the reversible steam reforming system. Chen and his coworkers have shown theoretically that the steam reforming rate of heptane is non-monotonic with respect to the partial pressure of hydrogen (Chen et al., 2003a). That is, the steam reforming rate of heptane increases when the partial pressure of hydrogen increases from 0 to 25.3kPa and then decreases after 25.3kPa.

At the entrance of the reformer, the partial pressure of hydrogen is usually smaller than 25.3kpa. But the steam reforming of heptane is a fast reaction at high temperature and pressure, which supplies a lot of hydrogen near the entrance of the reformer. As a result the partial pressure of hydrogen increases quickly. Although the removal of hydrogen decreases the partial pressure of hydrogen in the reaction side, the partial pressure of hydrogen is still higher than 25.3kPa due to the continuous fast production of hydrogen at high reaction temperatures and pressures. The removal of hydrogen increases the steam reforming rate of heptane and decreases the partial pressure of heptane (or the carbon formation rate from heptane decreases) at high reaction temperatures and pressures, leading to a higher reaction rate ratio between steam reforming of heptane and carbon formation from heptane cracking. Therefore, the amount of carbon formed from heptane for the case with hydrogen selective membranes is smaller than the case without hydrogen membranes at high reaction temperatures (>723K) and pressures (>1317kPa).

Secondly, the removal of hydrogen shifts the reversible steam reforming of methane and water gas shift reaction to the direction for hydrogen production. Because methane and carbon monoxide are the alternative carbon formation sources in the heptane steam reforming system, the shift of these reversible reactions to hydrogen production also makes the concentrations of methane and carbon monoxide much lower than the case without hydrogen selective membranes. As a result it suppresses the carbon formation from these two reforming by-products methane and carbon monoxide. Although it may also shifts the carbon formation by methane cracking at certain extent, the amount of carbon formed from methane is relatively smaller than the case without hydrogen selective membranes. Then the necessary amount of steam for the carbon complete gasification is smaller. Thus the critical S/C feed ratio for the case with hydrogen selective membranes is smaller than the case without hydrogen selective membranes at high reaction temperatures (>723K) and high pressures (>1317kPa), which is shown in Fig. 16.

Catalyst Deactivation and Engineering Control for Steam... 109

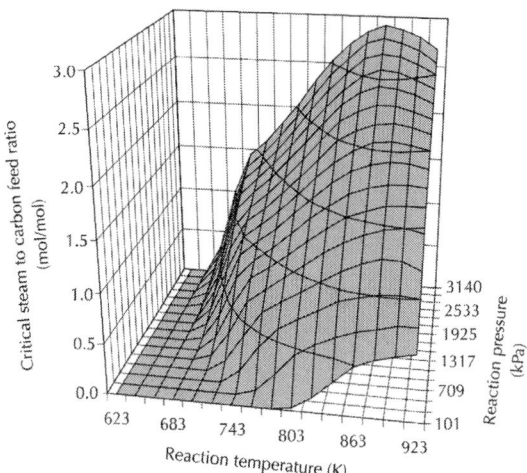

Figure 14: Carbon deposition free boundary for the case without hydrogen membranes.

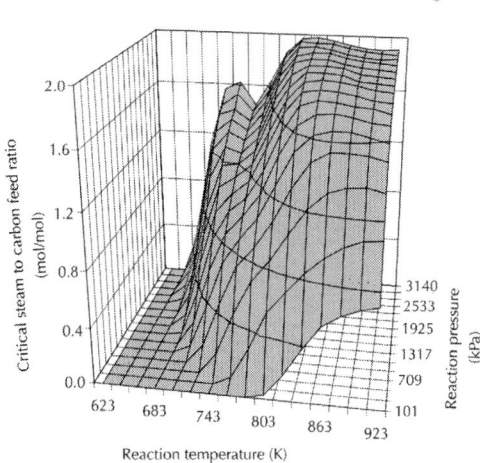

Figure 15: Carbon deposition free boundary for the case with 20 hydrogen membranes.

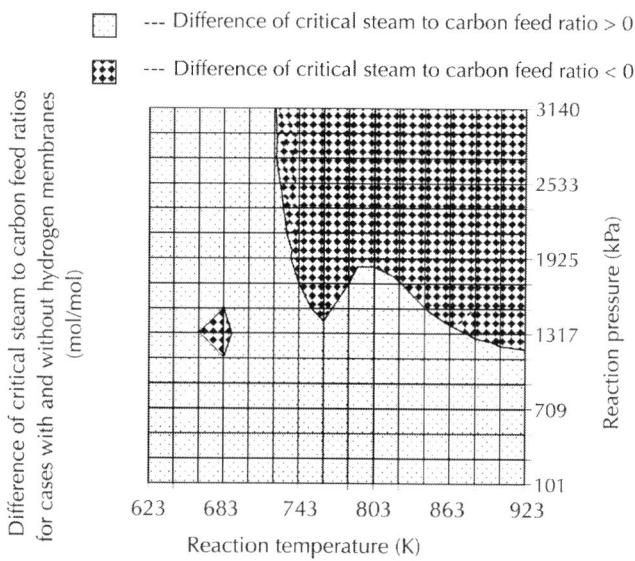

Figure 16: Difference of critical steam to carbon feed ratio for the cases with and without hydrogen selective membranes (difference of critical steam to carbon feed ratio=critical steam to carbon feed ratio for the case with hydrogen membranes−critical steam to carbon feed ratio for the case without hydrogen membranes).

As shown above it is possible to use these findings in the present investigation to guide the practical operation for the novel CFBMR regarding the carbon formation and catalyst deactivation. Although the reformer configuration is quite different from the other previous steam reformers, it is still possible to use the reported industrial/experimental data to check these findings. Table 4 shows some examples of comparison between the reported data and the model simulation. The critical S/C feed ratios in the last column of Table 4 are the model simulation results. They are obtained at the conditions with same reaction temperature and pressure for the reported industrial/experimental data. The reaction temperature is usually the average temperature from these reported data since there are temperature profiles in industrial or experimental reactors for the highly endothermic steam reforming. If no such information reported, the high temperature is usually used. The lowest operating

S/C feed ratios from the reported literatures can be regarded as the values close to the minimum S/C feed ratios. From the comparison of the last two columns in Table 4, it seems that the critical S/C feed ratio predicted from the model simulation agrees well with the reported industrial/experimental data. For example, Phillips 1969 and Phillips 1970 reported their experimental S/C feed ratio is about 1.429mol/mol at 14.7atm and 773K, which is the necessary ratio for satisfactory continuous operation of the industrial process. This necessary S/C feed ratio of 1.429mol/mol is very close to the model predicted critical S/C feed ratio of 1.251mol/mol.

Table 4: Examples of comparison between the reported industrial/experimental data and the model simulation

Reference	Hydrocarbon feed	Temperature (K)	Pressure (atm)	S/C feed ratio (mol/mol)	Critical S/C feed ratio by model simulation
Phillips 1969 and Phillips 1970	n-Heptane	773	14.7	1.429	1.251
Rostrup-Nielsen (1977)	Naphtha	~800	24.9	2.4–3.0	2.17
Tottrup (1982)	n-Heptane	773	20	1.9–7.4	1.57
Christensen (1996)	Naphtha	773	26	2.5–4.0	1.91

CONCLUSIONS

The nickel catalyst deactivation and CFBMR performance during steam reforming of heptane are investigated using an overall mathematical model including a random carbon deposition and catalyst deactivation model. Palladium based hydrogen selective membranes are used for the removal of product hydrogen, which "breaks" the thermodynamic equilibrium limitations associated with the reversible reforming reactions. As a result the yield of hydrogen with hydrogen selective membranes is much higher than the case without hydrogen selective membranes. The simulations show that the steam reforming of heptane has a strong tendency for carbon formation and deposition at low steam to carbon feed ratios (<1.4mol/mol) for high reaction temperatures (>700K) and pressures (>506.5kPa), which tends to deactivate the nickel reforming catalyst significantly. The effects of hydrogen selective membranes on the carbon deposition and catalyst deactivation are also investigated and the results are similar for both cases without and with hydrogen selective membranes. The catalyst activity decreases when steam to carbon feed ratio decreases, reaction temperature increases or reaction pressure increases. An engineering control approach, in-site control with a concept of critical/minimum steam to carbon feed ratio is suggested and used to determine the carbon deposition free boundary for both cases without and with hydrogen selective membranes in the CFBMR. It is found that at low reaction temperatures 623–723K and pressures 101–1317kPa, the critical steam to carbon feed ratios for the case with hydrogen selective membranes are higher than the case without hydrogen selective membranes. While at the other conditions where reaction temperature is higher than 723K and pressure is higher than 1317kPa, the critical steam to carbon feed ratios for the case with hydrogen selective membranes are smaller than the case without hydrogen selective membranes. The comparison between the reported data and model simulation shows that the critical S/C feed ratio predicted from the model agrees well with the reported industrial/experimental operating data. Thus it is possible to use these findings in the present investigation to guide the practical operation for the novel CFBMR regarding the carbon formation and catalyst deactivation as well as for other steam reformers.

ACKNOWLEDGMENTS

This work was financially supported by Auburn University, Grant Number 2-12085.

REFERENCES

1. Adris, A.M., Lim, C.J., Grace, J.R., 1994. The 2uidized bed membrane reactor (FBMR) system: a pilot scale experimental study. Chemical Engineering Science 49, 5833–5843.
2. Armor, J.N., 1999. Review: The multiple roles for catalysis in the production of H2. Applied Catalysis A: General 176, 159–176.
3. Barbieri, G., Di Maio, F.P., 1997. Simulation of methane steam reforming process in a catalytic Pd-membrane reactor. Industrial and Engineering Chemistry Research 36, 2121–2127.
4. Bartholomew, C.H., 2001. Mechanisms of catalyst deactivation. Applied catalysis A: General 212, 17–60.
5. Biswas, J., Do, D.D., 1987. A uniKed theory of coking deactivation in a catalyst pellet. Chemical Engineering Journal 36, 175–191.
6. Borowiecki, T., 1987. Nickel catalysts for steam reforming of hydrocarbons: direct and indirect factors alecting the coking rate. Applied Catalysis 31, 207–220.
7. Borowiecki, T., StasiYnska, B., Go lebiowski, A., 1997. Elects of small MoO3 additions on the properties of nickel catalysts for the steam reforming of hydrocarbons. Applied Catalysis A: General 141–156.
8. Chen, Z., Elnashaie, S.S.E.H., 2002. EScientproduction of hydrogen from higher hydrocarbons using novel membrane reformer. 14th World Hydrogen Energy Conference, Montreal, Canada, June 9–13.
9. Chen, C.X., Masayuki Horio, Toshinori Kojima, 2000. Numerical simulation of entrained 2ow coal gasiKers. Part I: modeling of coal gasiKcation in an entrained 2ow gasiKer. Chemical Engineering Science 55, 3861–3874.
10. Chen, Z., Prasad, P., Elnashaie, S.S.E.H., 2002. The coupling of catalytic steam reforming and oxidative reforming of

methane to produce pure hydrogen in a novel circulating fast 2uidized bed membrane reformer. ACS Meeting, Orlando, FL, Fuel Chemistry Division Preprints 47 (1), 111–113.
11. Chen, Z., Yan, Y., Elnashaie, S.S.E.H., 2003a. Modeling and optimization of a novel membrane reformer for higher hydrocarbons. A.I.Ch.E. Journal 49 (5), 1250–1265.
12. Chen, Z., Yan, Y., Elnashaie, S.S.E.H., 2003b. Non-monotonic behavior of hydrogen production from higher hydrocarbon steam reforming in a circulating fast 2uidized bed membrane reformer. Industrial and Engineering Chemistry Research 42 (25), 6549–6558.
13. Christensen, T.S., 1996. Adiabatic prereforming of hydrocarbons—an important step in syngas production. Applied Catalysis A: General 138, 285–309.
14. Elnashaie, S.S.E.H., Elshishini, S.S., 1993. Modelling, Simulation and Optimization of Industrial Fixed Bed Catalytic Reactors. Gordon and Breach Science Publishers, London, UK.
15. El Solh, T., Jarosch, K., de Lasa, H.I., 2001. Fluidized catalyst for methane reforming. Applied Catalysis A: General 210 (1–2), 315–324.
16. Forzatti, P., Lietti, L., 1999. Catalyst deactivation. Catalysis Today 52, 165–181.
17. Goltsov Victor, A., Nejat Veziroglu, T., 2002. A step on the road to hydrogen civilization. International Journal of Hydrogen Energy 27 (7–8), 719–723.
18. Kepinski, L., Stasinska, B., Borowiecki, T., 2000. Carbon deposition on Ni=Al2O3 catalysts doped with small amounts of Molybdenum. Carbon 38, 184–185.
19. Kunii, D., Levenspiel, O., 1990. Entrainment of solids from 2uidized beds: I. Hold-up of solids in the freeboard, II. Operation of fast 2uidized beds. Powder Technology 61, 193–206.
20. Kunii, D., Levenspiel, O., 1997. Circulating 2uidized-bed reactors. Chemical Engineering Science 15, 2471–2484.

21. Ohi, J., 2002. Hydrogen energy futures: scenario planning by the U.S. DOE hydrogen technical advisory panel. 14th World Hydrogen Energy Conference, Montreal, Canada, June 9–13.
22. Olsbye, U., Moen, O., Slagtern, A., Dahl, I.M., 2002. An investigation U of the coking properties of Kxed and 2uid bed reactors during methane-to-synthesis gas reactions. Applied Catalysis A: General 228, 289–303.
23. Perry, R.H., Chilton, C.H., Kirkpatrick, S.D., 1984. Chemical Engineers' Handbook, 6th Edition. Mcgraw-Hill Book Co., New York.
24. Phillips, T.R., Mulhall, J., Turner, G.F., 1969. The kinetics and mechanism of the reaction between steam and hydrocarbons over Nickel catalysts in the temperature range 350–500°C, PartI. Journal of Catalysis 15, 233.
25. Phillips, T.R., Mulhall, J., Turner, G.F., 1970. The kinetics and mechanism of the reaction between steam and hydrocarbons over Nickel catalysts in the temperature range 350–500°C, PartII. Journal of Catalysis 17, 28.
26. Ren, X.-H., Bertmer, M., Stapf, S., Demco, D.E., Blumich, B., Kern, C., \ Jess, A., 2002. Deactivation and regeneration of a naphtha reforming catalyst. Applied Catalysis A: General 39–52.
27. Rostrup-Nielsen, J.R., 1974. Coking on Nickel catalysts for steam reforming of hydrocarbons. Journal of Catalysis 33, 184–201.
28. Rostrup-Nielsen, J., 1977. Hydrogen via steam reforming of Naphtha. Chemical Engineering Progress 9, 87.
29. Rostrup-Nielsen, J.R., 1979. Symposium on the science of catalysis and its application in industry, Sindri, India, 22–24.
30. Rostrup-Nielsen, J.R., 1997. Industrial relevance of coking. Catalysis Today 37, 225–232.
31. Scholz, W.H., 1993. Processes for industrial production of hydrogen and associated environmental elects. Gas Separation and PuriKcation 7, 131–139.

32. Sehested, J., Carlsson, A., Janssens, T.V.W., Hansen, P.L., Datye, A.K., 2001. Sintering of Nickel steam-reforming catalysts on MgAl2O4 spinel supports. Journal of Catalysis 197, 200–209.
33. Shu, B.P.A.G., Kaliaguine, S., 1994. Methane steam reforming in asymmetric Pd- and Pd-Ag/porous SS membrane reactor. Applied Catalysis A 119, 305–325.
34. Snoeck, J.W., Froment, G.F., Fowles, M., 1997. Kinetic study of the carbon Klament formation by methane cracking on a Nickel catalyst. Journal of Catalysis 169, 250–262.
35. Tottrup, P.B., 1976. Kinetics of decomposition of carbon monoxide on a supported Nickel catalyst. Journal of Catalysis 42, 29–36.
36. Tottrup, P.B., 1982. Evaluation of intrinsic steam reforming kinetic parameters from rate measurements on full particle size. Applied Catalysis 4, 377–389.
37. Trimm, D.L., 1984. Control of coking. Chemical Engineering Process 18, 137–148.
38. Trimm, D.L., 1999. Catalysts for the control of coking during steam reforming. Catalysis Today 49, 3–10.
39. Twigg, M.V., 1989. Catalyst Handbook, 2nd Edition, Wolfe Publishing Ltd, England, pp. 225–282.
40. Verykios, X.E., 2003. Catalytic dry reforming of natural gas for the production of chemicals and hydrogen. International Journal of Hydrogen Energy 28 (10), 1045–1063.
41. Vooehies Jr., A., 1945. Industrial and Engineering Chemistry 37, 318.
42. Xu, J., Froment, G.F., 1989. Froment, Methane steam reforming, methanation and water–gas shift: I. Intrinsic kinetics. Journal of AIChE 35 (1), 88–96.

Chapter 4

Alternate Strategies for Conversion of Waste Plastic to Fuels

Neha Patni, Pallav Shah, Shruti Agarwal, and Piyush Singhal

Department of Chemical Engineering, Institute of Technology, Nirma University, S. G. Highway, Ahmedabad, Gujarat 382481, India

ABSTRACT

The present rate of economic growth is unsustainable without saving of fossil energy like crude oil, natural gas, or coal. There are many alternatives to fossil energy such as biomass, hydropower, and wind energy. Also, suitable waste management strategy is another important aspect. Development and modernization have brought about a huge increase in the production of all kinds of commodities, which indirectly generate waste. Plastics have been one of the materials because of their wide range of applications due to versatility and relatively low cost. The paper presents the current scenario of the plastic consumption. The aim is to provide the reader with an in depth analysis regarding the recycling techniques of plastic solid waste (PSW). Recycling can be divided into four categories: primary, secondary, tertiary, and quaternary. As calorific value of the plastics is comparable to that of fuel, so production of fuel would be a better alternative. So the methods of converting plastic into fuel, specially pyrolysis and catalytic degradation, are discussed in detail and a brief idea about the gasification is also included. Thus, we attempt to address the problem of plastic waste disposal and shortage of conventional fuel and thereby help in promotion of sustainable environment.

INTRODUCTION

The increase in use of plastic products caused by sudden growth in living standards had a remarkable impact on the environment. Plastics have now become indispensable materials, and the demand is continually increasing due to their diverse and attractive applications in household and industries. Mostly, thermoplastics polymers make up a high proportion of waste, and this amount is continuously increasing around the globe. Hence, waste plastics pose a very serious environmental challenge because of their huge quantity and disposal problem as thermoplastics do not biodegrade for a very long time.

The consumption of plastic materials is vast and has been growing steadily in view of the advantages derived from their versatility, relatively low cost, and durability (due to their high chemical stability and low degradability). Some of the most used plastics are polyolefins such as polyethylene and polypropylene, which have a massive production and consumption in many applications such as packaging, building, electricity and electronics, agriculture, and health care [1]. In turn, the property of high durability makes the disposal of waste plastics a very serious environmental problem, land filling being the most used disposal route. Plastic wastes can be classified as industrial and municipal plastic wastes according to their origins; these groups have different qualities and properties and are subjected to different management strategies [2,3].

Plastic materials production has reached global maximum capacities leveling at 260 million tons in 2007, where in 1990 the global production capacity was estimated at 80 million tons [1]. Plastic production is estimated to grow worldwide at a rate of about 5% per year [4]. Polymer waste can be used as a potentially cheap source of chemicals and energy. Due to release of harmful gases like dioxins, hydrogen chloride, airborne particles, and carbon dioxide, incineration of polymer possesses serious air pollution problems. Due to high cost and poor biodegradability, it is also undesirable to dispose by landfill.

Recycling is the best possible solution to the environmental challenges facing the plastic industry. These are categorized into primary, secondary, tertiary, and quaternary recycling. Chemical recycling, that is, conversion of waste plastics into feedstock or fuel has been recognized as an ideal approach and could significantly reduce the net cost of disposal. The production of liquid hydrocarbons from plastic degradation would be beneficial in that liquids are easily stored, handled, and transported. However, these aims are not easy to achieve [4]. An alternative strategy to chemical recycling, which has attracted much interest recently, with the aim of converting waste plastics into basic petrochemicals is to be used as hydrocarbon feedstock or fuel oil for a variety of downstream processes [3]. There are different methods of obtaining fuel from waste plastic such as thermal degradation, catalytic cracking, and gasification [3, 5].

CURRENT SCENARIO OF PLASTICS

Over many years, a drastic growth has been observed in plastic industry such as in the production of synthetic polymers represented by polyethylene (PE), polypropylene (PP), polystyrene (PS), polyethylene terephthalate (PET), polyvinyl alcohol (PVA), and polyvinyl chloride (PVC). It has been estimated that almost 60% of plastic solid waste (PSW) is discarded in open space or land filled worldwide. According to a nationwide Survey conducted in the year 2003, more than 10,000 MT of plastic waste is generated daily in our country, and only 40 wt% of the same is recycled; balance 60 wt% is not possible to dispose off [4]. India has been a favored dumping ground for plastic waste mostly from industrialized countries like Canada, Denmark, Germany, U.K, the Netherlands, Japan, France, and the United States of America. According to the government of India, import data of more than 59,000 tons and 61,000 tons of plastic waste have found its way into India in the years 1999 and 2000, respectively [3, 6].

Present Scenario in India

With the formal and informal sector failing to collect plastic waste the packaging and polyvinyl chloride (PVC) pipe industry are growing at 16–18% per year. The demand of plastic goods is increasing from household use to industrial applications. It is growing at a rate of 22% annually. The polymers production has reached the 8.5 million tons in 2007. Table 1 provides the total plastics waste consumption in the world and Table 2 provides the total plastic waste consumption in India during the last decade. National plastic waste management task force in 1997 projected the polymers demand in the country. Table 3 documents the demand of different polymers in India during years 1995-96, 2001-02, and 2006-07. The comparison of demand and consumption from Tables 2 and 3 indicates that projections are correct. More than one fourth of the consumption in India is that of PVC, which is being phased out in many countries. Poly bags and other plastic items except PET in particular have been a focus, because it has contributed to host problems in India such as choked sewers, animal deaths, and clogged soils.

Table 1: Plastics consumption, by major world areas, in kg and GNI dollars per capita

Main world areas	Plastics consumption, 000s tons	Population millions	Kg/capita	GNI/capita
Europe W, C, and E	40 000	450	90	18 000
Eurasia, Russia, and others	4 000	285	14	1 600
North America	45 000	310	145	32 000
Latin America	11 000	500	22	3 500
Middle East, including TR	4 000	200	20	2 500
Africa, North and South	2 500	190	13	2 000
Other Africa	500	610	<1	300
China	19 000	1285	14	800
India	4 000	1025	4	450
Japan	11 000	125	90	35 000
Other Asia Pacific, rest	13 000	1120	11	600
Total world	154 000	6 100	25	5 200

Table 2: Plastics consumption in India

S. no.	Year	Consumption (tons)
1	1996	61,000
2	2000	3,00,000
3	2001	4,00,000
4	2007	85,00,000

Table 3: Polymers demands in India (million tons)

S. no	Type of polymer	1995-96	2001-02	2006-07
1	Polyethylene	0.83	1.83	3.27
2	Polypropylene	0.34	0.88	1.79
3	Polyvinyl chloride	0.49	0.87	1.29
4	Polyethylene terephthalate	0.03	0.14	0.29

Source: National Plastic Waste Management Task Force Projection (1997).

DIFFERENT RECYCLING CATEGORIES [1]

Primary Recycling

It is also known as mechanical reprocessing. During the process, the plastic waste is fed into the original production process of basic material. So, we can obtain the product with same specification as that of the original one. This process is feasible only with semiclean scrap, so it is an unpopular choice with the recyclers. Degraded plastic waste partly substitutes the virgin material. So, on increasing the recycled plastic fraction in feed mixture, the quality of the product decreases. This type of recycling requires clean and not contaminated waste which is of the same type as virgin resin.

For this reason, steps in the primary recycling process are:
- separate the waste by specific type of resin and by different colors and then wash it,
- the waste has better melting properties so it should be reextruded into pellets which can be added to the original resin.This type of recycling is very expensive compared to other types of recycling due to the requirements of plastic properties mentioned above.

If the waste can be easily sorted by resin but cannot be pelletized due to mixed coloring contamination, then waste can be fed into moulding application, and regarding reactants properties, it is less demanding.

Secondary Recycling

Secondary recycling uses PSW in the manufacturing of plastic products by mechanical means, which uses recyclates, fillers, and/or virgin polymers. The objective of the process is to retain some energy which is used for plastic production to attain financial advantages. Unlike primary recycling, the secondary recycling process can use contaminated or less separated waste. However, this waste has to be cleaned. The recycling process involves different products and is different compared to original production process.

Tertiary Recycling

This process is also known as cracking process. The process includes breaking down the plastics at high temperatures (thermal degradation) or at lower temperatures in the presence of catalyst (catalytic degradation), which contain smaller carbon chains. For any chemical production, this feedstock can be used as basic material of lower quality (e.g., polymerization or fuel fabrication). The original value of the raw material is lost. The tertiary recycling process is more important due to high levels of waste contamination. We are able to recover the monomers of condensation polymers. Mechanisms like hydrolysis, methanolysis, or glycolysis can be used, for example, PET (polyethylene terephthalate), polyesters, and polyamide while addition of polymers like polyolefin, polystyrene, and PVC requires stronger thermal treatment, gasification, or catalytic degradation to be cracked.

Quaternary Recycling

This process includes the recovery of energy content only. As most plastic waste has high heat content so it is incinerated. Generation of the heat energy is the only advantage of this process. The residual of this incineration has 20wt%, respectively, 10 vol% of the original waste and are placed in landfills. Solid waste problem is not solved by this process; in fact it leads to the problem of air pollution.

METHODS OF CONVERTING PLASTIC TO FUEL

Pyrolysis/Thermal Degradation

Pyrolysis is a process of thermal degradation of a material in the absence of oxygen. Plastic is fed into a cylindrical chamber. The pyrolytic gases are condensed in a specially designed condenser system, to yield a hydrocarbon distillate comprising straight and branched chain aliphatic, cyclic aliphatic, and aromatic hydrocarbons, and liquid is separated using fractional distillation to produce the liquid fuel products. The plastic is pyrolysed at 370°C–420°C.

The essential steps in the pyrolysis of plastics involve (figure 1):
- evenly heating the plastic to a narrow temperature range without excessive temperature variations,
- purging oxygen from pyrolysis chamber,
- managing the carbonaceous char by-product before it acts as a thermal insulator and lowers the heat transfer to the plastic,
- careful condensation and fractionation of the pyrolysis vapors to produce distillate of good quality and consistency.

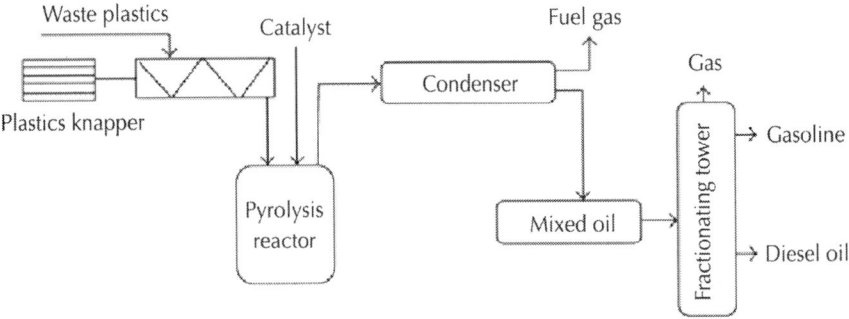

Figure 1: Pyrolysis Process of generating fuel oil from the waste plastics [12].

Advantages of pyrolysis process [5] are
- volume of the waste is significantly reduced (<50–90%),
- solid, liquid, and gaseous fuel can be produced from the waste,
- storable/transportable fuel or chemical feed stock is obtained,
- environmental problem is reduced,
- desirable process as energy is obtained from renewable sources like municipal solid waste or sewage sludge,
- the capital cost is low.

There are different types of pyrolysis process. Conventional pyrolysis (slow pyrolysis) proceeds under a low heating rate with solid, liquid, and gaseous products in significant portions [5, 13]. It is an ancient process used mainly for charcoal production. Vapors can be continuously removed as they are formed [5, 14]. The fast pyrolysis is associated with tar, at low temperature (850–1250 K) and/or gas at high temperature (1050–1300 K). At present, the preferred technology is fast or flash pyrolysis at high temperatures with very short residence time [5, 15]. Fast pyrolysis (more accurately defined as thermolysis) is a process in which a material, such as biomass, is rapidly heated to high temperatures in the absence of oxygen [5, 15]. Table 4 [7] shows the range of the main operating parameters for pyrolysis processes.

Table 4: Main operating parameters for pyrolysis process [7]

Parameters	Conventional	Fast	Flash
Pyrolysis temprature (K)	550–900	850–1250	1050–1300
Heating rate (K/s)	0.1–1	10–200	>1000
Particle size (mm)	5–50	<1	<0.2
Solid residence (s)	300–3600	0.5–10	<0.5

Mechanism of Thermal Degradation

Cullis and Hirschler had proposed detailed study on the mechanism of thermal degradation of polymers [3, 16]. The four different mechanisms proposed are: (1) end-chain scission or unzipping, (2) random-chain scission/fragmentation, (3) chain stripping/elimination of side chain, (4) cross-linking. The decomposition mode mainly depends on the type of polymer (the molecular structure):

$$M_n^* \longrightarrow M_{n-1}^* + M \quad (1)$$

$$M_{n-1}^* \longrightarrow M_{n-2}^* + M \quad (2)$$

$$M_n \longrightarrow M_x + M_y \quad (3)$$

Equations (1) and (2) represent the thermal degradation, and (3) represents the random degradation route of the polymers pyrolysis. The fourth type of mechanism, that is, cross-linking often occurs in thermosetting plastics upon heating at high temperature in which two adjacent "stripped" polymer chains can form a bond resulting in a chain network (a higher MW species). An example is char formation.

Catalytic Degradation

In this method, a suitable catalyst is used to carry out the cracking reaction. The presence of catalyst lowers the reaction temperature and time. The process results in much narrower product distribution of carbon atom number and peak at lighter hydrocarbons which occurs at lower temperatures. The cost should be further reduced to make the process more attractive from an economic perspective. Reuse of catalysts and the use of effective catalysts in lesser quantities can optimize this option. This process can be developed into a cost-effective commercial polymer recycling process for solving the acute environmental problem of disposal of plastic waste. It also offers the higher cracking ability of plastics, and the lower concentration of solid residue in the product [3].

Mechanism of Catalytic Degradation

Singh et al. [3, 17] have investigated catalytic degradation of polyolefin using TGA as a potential method for screening catalysts and have found that the presence of catalyst led to the decrease in the apparent activation energy. Different mechanisms (ionic and free radical) for plastic pyrolysis proposed by different scientists are as given below.

There are different steps in carbonium ion reaction mechanism such as H-transfer, chain/beta-scission, isomerisation, oligomerization/alkylation, and aromatization which is influenced by acid site strength, density, and distribution [3, 18]. Solid acid catalysts, such as zeolites, favor hydrogen transfer reactions due to the presence of many acid sites [3, 5]. both Bronsted and Lewis acid sites characterize acid strength of solid acids. The presence of Bronsted acid sites supports the cracking of olefinic compounds [3, 19].The majority of the acid sites in crystalline solid acids are located within the pores of the material, such as with zeolites [3, 20]. Thus, main feature in assessing the level of polyolefin cracking over such catalysts is the microporosity of porous solid acids. The carbonium ion mechanism of catalytic pyrolysis of polyethylene can be described as follows [3, 21] (see Table 5).

Table 5: List of catalysts in use

Sr. no.	Catalyst	Pore size (nm)	Commercial name	References
1	USY	0.74	H-Ultrastabilised, Y-zeolite	[8–10]
2	ZSM-5	0.55 × 0.51	H-ZSM-5 zeolite	[8–10]
3	MOR	0.65 × 0.70	H-Mordenite	[8, 9]
4	ASA	3.15	Synclyst 25 (silica-alumina)	[8, 9]
5	MCM-41	4.2–5.2	—	[8, 9, 11]
6	SAHA	3.28	Amorphous silica-alumina	[10]
7	FCC-R1	—	Equilibrium catalyst	[10]
8	Silicalite	0.55 × 0.51	Synthesized In house	[10]

(1) Initiation. Initiation may occur on some defected sites of the polymer chains. For instance, an olefinic linkage could be converted into an on-chain carbonium ion by proton addition:

$$-CH_2CH_2CH = CHCH_2CH_2^- + HX$$
$$\longrightarrow CH_2CH_2^+CHCH_2-CH_2CH_2 + X^- \qquad (4)$$

The polymer chain may be broken up through β-emission:

$$-CH_2CH_2^+CHCH_2-CH_2CH_2^-$$
$$\longrightarrow CH_2CH_2CH = CH_2 + {}^+CH_2CH_2^+ \qquad (5)$$

Initiation may also take place through random hydride-ion abstraction by low-molecular-weight carbonium ions (R^+):

$$-CH_2CH_2CH_2CH_2CH_2- + R^+$$
$$\longrightarrow -CHCH_2^+CHCH_2CH_2^- + RH \qquad (6)$$

The newly formed on-chain carbonium ion then undergoes -scission.

(2) Depropagation. The molecular weight of the main polymer chains may be reduced through successive attacks by acidic sites or other carbonium ions and chain cleavage, yielding ingan oligomer fraction (approximately C_{30}–C_{80}). FURTHER,

cleavage of the oligomer fraction probably by direct -emission of chain-end carbonium ions leads to gas formation on one hand and a liquid fraction (approximately C_{10}–C_{25}) on the other.

(3) Isomerization. The carbonium ion intermediates can undergo rearrangement by hydrogen- or carbon-atom shifts, leading to a double-bond isomerization of an olefin:

$$CH_2 = CH-CH_2-CH_2-CH_3$$

$$\xrightarrow{H^+} CH_3{}^+CH-CH_2-CH_2-CH_3$$

$$\xrightarrow{H^+} CH_3-CH = CH-CH_2-CH_3 \qquad (7)$$

Other important isomerization reactions are methyl-group shift and isomerization of saturated hydrocarbons.

(4) Aromatization. Some carbonium ion intermediates can undergo cyclization reactions. An example is when hydride ion abstraction first takes place on an olefin at a position several carbons removed from the double bond, the result being the formation of an olefinic carbonium ion:

$$R_1^+ + R_2CH = CH-CH_2CH_2CH_2CH_2CH_3$$

$$\longleftrightarrow R_1H + R_2CH = CH-CH_2CH_2CH_2{}^+CHCH_3 \qquad (8)$$

The carbonium ion could undergo intramolecular attack on the double bond.

Panda et al. [3] and Sekine, and Fujimoto [22] have proposed a free radical mechanism for the catalytic degradation of PP using Fe/activated carbon catalyst. Methyl, primary and secondary alkyl radicals are formed during degradation and methane, olefins and monomers are produced by hydrogen abstractions and recombination of radical units [3, 23].

The various steps in catalytic degradation are shown below [3].

(1) Initiation. Random breakage of the C–C bond of the main chain occurs with heat to produce hydrocarbon radicals:

$$R_1-R_2 \longrightarrow R_1^\bullet + R_2^\bullet \qquad (9)$$

(2) Propagation. The hydrocarbon radical decomposes to produce lower hydrocarbons such as propylene, followed by -scission and abstraction of H-radicals from other hydrocarbons to produce a new hydrocarbon radical:

$$R_1^{\bullet} \longrightarrow R_3^{\bullet} + C_2 \text{ or } C_3 \qquad (10)$$

$$R_2^{\bullet} + R_4 \longrightarrow R_2 \text{ or } R_4^{\bullet} \qquad (11)$$

(3) Termination. Disproportionation or recombination of two radicals:

$$R_5^{\bullet} + R_6^{\bullet} \longrightarrow R_5 + R_6^{\bullet} \qquad (12)$$

$$R_7^{\bullet} + R_8^{\bullet} \longrightarrow R_7 - R_8 \qquad (13)$$

During catalytic degradation with Fe activated charcoal in H_2 atmosphere, hydrogenation of hydrocarbon radical (olefin) and the abstraction of the H-radical from hydrocarbon or hydrocarbon radical generate radicals, and thus, enhancing degradation rate. At reaction temperature lower than 400°C or a reaction time shorter than 1.0 h, many macromolecular hydrocarbon radicals exist in the reactor, and recombination occurs readily because these radicals cannot move fast. However, with Fe activated carbon in a H_2 atmosphere, these radicals are hydrogenated, and therefore, combination may be suppressed. Consequently, it seems as if the decomposition of the solid product is promoted, including low polymers whose molecular diameter is larger than the pore size of the catalysts.

Gasification

In this process, partial combustion of biomass is carried out to produce gas and char at the first stage and subsequent reduction of the product gases, chiefly CO_2 and H_2O, by the charcoal into CO and H_2. Depending on the design and operating conditions of the reactor, the process also generates some methane and other higher hydrocarbons (HCs) [5, 24]. Broadly, gasification can be defined as the thermochemical conversion of a solid or liquid carbon-based material (feedstock) into a combustible gaseous product (combustible gas) by the supply of a gasification agent (another gaseous compound). The gasification agent allows the feedstock to be quickly converted into gas by means of different heterogeneous reactions [5,25–27]. If the process does not occur with help of an oxidising agent, it is called indirect gasification and needs an external energy source gasification agent, because it is easily produced and increases the hydrogen content of the combustible gas [5, 28].

A gasification system is made up of three fundamental elements: (1) the gasifier, helpful in producing the combustible gas; (2) the gas clean up system, required to remove harmful compounds from the combustible gas; (3) the energy recovery system. The system is completed with suitable subsystems, helpful to control environmental impacts (air pollution, solid wastes production, and wastewater).

Gasification process represents a future alternative to the waste incinerator for the thermal treatment of homogeneous carbon based waste and for pretreated heterogeneous waste.

SUMMARY

Plastics are "one of the greatest innovations of the millennium" and have certainly proved their reputation to be true. Plastic is lightweight, does not rust or rot, is of low cost, reusable, and conserves natural resources and for these reasons, plastic has gained this much popularity. The literature reveals that research efforts on the pyrolysis of plastics in different conditions using different catalysts and the process have been initiated. However, there are many subsequent problems to be solved in the near future. The present issues are the necessary scale up, minimization of waste handling costs and production cost, and optimization of gasoline range products for a wide range of plastic mixtures or waste. Huge amount of plastic wastes produced may be treated with suitably designed method to produce fossil fuel substitutes. The method is superior in all respects (ecological and economical) if proper infrastructure and financial support is provided. So, a suitable process which can convert waste plastic to hydrocarbon fuel is designed and if implemented then that would be a cheaper partial substitute of the petroleum without emitting any pollutants. It would also take care of hazardous plastic waste and reduce the import of crude oil.

Challenge is to develop the standards for process and products of postconsumer recycled plastics and to adopt the more advanced pyrolysis technologies for waste plastics, referring to the observations of research and development in this field. The pyrolysis reactor must be designed to suit the mixed waste plastics and small-scaled and middle-scaled production. Also, analysis would help reducing the capital investment and also the operating cost and thus would enhance the economic viability of the process.

REFERENCES

1. T. S. Kpere-Daibo, Plastic catalytic degradation study of the role of external catalytic surface, catalytic reusability and temperature effects [Doctoral thesis], University of London department of Chemical Engineering University College London, WC1E 7JE.
2. A. G. Buekens and H. Huang, "Catalytic plastics cracking for recovery of gasoline-range hydrocarbons from municipal plastic wastes," Resources Conservation and Recycling, vol. 23, no. 3, pp. 163–181, 1998.·
3. A. K. Panda, R. K. Singh, and D. K. Mishra, "Thermolysis of waste plastics to liquid fuel. A suitable method for plastic waste management and manufacture of value added products—a world prospective," Renewable and Sustainable Energy Reviews, vol. 14, no. 1, pp. 233–248, 2010. ·

4. S. M. Al-Salem, P. Lettieri, and J. Baeyens, "The valorization of plastic solid waste (PSW) by primary to quaternary routes: from re-use to energy and chemicals," Progress in Energy and Combustion Science, vol. 36, no. 1, pp. 103–129, 2010. ·
5. R. P. Singhad, V. V. Tyagib, T. Allen, et al., "An overview for exploring the possibilities of energy generation from municipal solid waste (MSW) in Indian scenario," Renewable and Sustainable Energy reviews, vol. 15, no. 9, pp. 4797–4808, 2011. ·
6. J. Scheirs and W. Kaminsky, Feedstock Recycling of Waste Plastics, John Wiley & Sons, 2006.
7. A. Demirbas, "Biorefineries: current activities and future DEVELOPMENTS," Energy Conversion & Management, vol. 50, pp. 2782–2801, 2009.
8. W.-C. Huang, M.-S. Huang, C.-F. Huang, C.-C. Chen, and K.-L. Ou, "Thermochemical conversion of polymer wastes into hydrocarbon fuels over various fluidizing cracking catalysts," Fuel, vol. 89, no. 9, pp. 2305–2316, 2010. ·
9. T.-T. Wei, K.-J. Wu, S.-L. Lee, and Y.-H. Lin, "Chemical recycling of post-consumer polymer waste over fluidizing cracking catalysts for producing chemicals and hydrocarbon fuels," Resources, Conservation and Recycling, vol. 54, no. 11, pp. 952–961, 2010. ·
10. H.-T. Lin, M.-S. Huang, J.-W. Luo, L.-H. Lin, C.-M. Lee, and K.-L. Ou, "Hydrocarbon fuels produced by catalytic pyrolysis of hospital plastic wastes in a fluidizing cracking process," Fuel Processing Technology, vol. 91, no. 11, pp. 1355–1363, 2010. ·
11. J. Aguado, D. P. Serrano, and J. M. Escola, "Fuels from waste plastics by thermal and catalytic process: a review," industrial & Engineering Chemistry Research, vol. 47, no. 21, pp. 7982–7992, 2008. ·
12. G. H. Zhang, J. F. Zhu, and A. Okuwaki, "Prospect and current status of recycling waste plastics and technology for converting them into oil in China," Resources, Conservation

and Recycling, vol. 50, no. 3, pp. 231–239, 2007. ·
13. S. Katyal, "effect of carbonization temperature on combustion reactivity ofbagasse char," Energy Sources A, vol. 29, no. 16, pp. 1477–1485, 2007. ·
14. D. Mohan, C. U. Pittman Jr., and P. H. Steele, "Pyrolysis of wood/biomass for bio-oil: acritical review,"Energy Fuels, vol. 20, no. 3, pp. 848–889, 2006. ·
15. A. Demirbas, "Producing bio-oil from olive cake by fast pyrolysis," Energy Sources A, vol. 30, pp. 38–44, 2008.
16. C. F. Cullis and M. M. Hirschler, The Combustion of Organic Polymers, Oxford Clarendon press, 1981.
17. B. Singh and N. Sharma, "Mechanistic implications of plastic degradation," Polymer Degradation and Stability, vol. 93, no. 3, pp. 561–584, 2008. ·
18. A. Corma, "Inorganic solid acids and their use in acid-catalyzed hydrocarbon reactions," Chemical Reviews, vol. 95, no. 3, pp. 559–614, 1995.
19. H. Ohkita, R. Nishiyama, Y. Tochihara et al., "Acid properties of silica-alumina catalysts and catalytic degradation of polyethylene," Industrial and Engineering Chemistry Research, vol. 32, no. 12, pp. 3112–3116, 1993.
20. P. Venuto and P. Landis, "Zeolite catalysis in synthetic organic chemistry," Advances in Catalysis, vol. 18, pp. 259–267, 1968. ·
21. A. G. Buekens and H. Huang, "Catalytic plastics cracking for recovery of gasoline-range hydrocarbons from municipal plastic wastes," Resources, Conservation and Recycling, vol. 23, no. 3, pp. 163–181, 1998.·
22. Y. Sekine and K. Fujimoto, "Catalytic degradation of PP with an Fe/activated carbon catalyst," Journal of material Cycles and Waste Management, vol. 5, no. 2, pp. 107–112, 2003. ·
23. R. P. Lattimer, "direct analysis of polypropylene compounds by thermal desorption and pyrolysis-mass spectrometry,"

Journal of Analytical and applied Pyrolysis, vol. 26, no. 2, pp. 65–92, 1993.

24. H. R. Appel, Y. C. Fu, S. Friedman, P. M. Yavorsky, and I. Wender, "Converting organic wastes to oil,"U.S. Burea of Mines Report of Investigation 7560, 1971.

25. C. Di Blasi, "Dynamic behaviour of stratified downdraft gasifier," Chemical Engineering Science, vol. 55, no. 15, pp. 2931–2944, 2000. ·

26. G. Barducci, "The RDF gasifier of Florentine area (Gréve in Chianti Italy)," in Proceedings of the 1st Italian-Brazilian Symposium on Sanitary and Environmental Engineering, 1992.

27. S. Z. Baykara and E. Bilgen, "A feasibility study on solar gasification of albertan coal," in Alternative Energy Sources IV, vol. 6, Ann Arbor Science, New York, NY, USA, 1981.

28. W. B. Hauserman, N. Giordano, M. Lagana, and V. Recupero, "Biomass gasifiers for fuelcells systems," La Chimica & L'Industria, vol. 2, pp. 199–206, 1997.

Chapter 5

Systematic Retrofit Design with Response Surface Method and Process Integration Techniques: A Case Study for the Retrofit of A Hydrocarbon Fractionation Plant

Aurora Hernández Enríquez[a], Michael Binns[b], and Jin-Kuk Kim[b]

[a]Centre for Process Integration, School of Chemical Engineering and Analytical Science, The University of Manchester, Manchester M13 9PL, UK

[b]Department of Chemical Engineering, Hanyang University, Wangsimni-ro 222, Seongdong-gu, Seoul 133-791, Republic of Korea

ABSTRACT

This paper demonstrates the Retrofit Design Approach (RDA) and Response Surface Methodology (RSM) for the retrofit of industrial plants in which assessment of design options for improving existing processes in a site-wide and integrated manner is not straightforward, due to complex design interactions in the process. The design methodology applied in this study is based on the systematic use of a process simulator which is used to identify promising variables through sensitivity analysis. Hence, the most important factors are determined and a reduced model is constructed based on RSM. An optimization framework is then built using the reduced model based on key selected variables, which is optimized to find optimal conditions and performance of the process. This design methodology provides strategic guidelines for determining the most cost-effective design options. The retrofit of a hydrocarbon fractionation plant is presented as an industrial case study. This includes a large number of design options with different process configurations and operating conditions due to the interconnection of distillation columns in sequence and the integrated heat recovery within the plant. The case study results demonstrate the applicability of the proposed approach which is able to effectively deal with a large retrofit problems. This is possible with the aid of process simulation and RSM producing a reduced model which requires considerably less computational effort to solve.

GRAPHICAL ABSTRACT

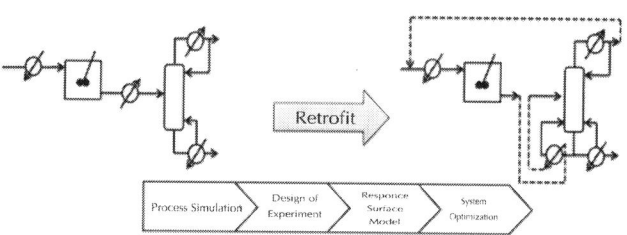

INTRODUCTION

In process industries, increasing profits is not the only consideration and maintaining a safe working environment and improving sustainability are also very important. Modern plants are designed and operated in a highly integrated manner and consequently techno-economic analysis of existing plants for debottlenecking or retrofitting is too complicated to be dealt with using a simple model. In order to identify cost-effective process changes and evaluate their impacts on the site, it is necessary to build a modelling and design framework which allows full and thorough investigation of the plants overall performance. For small industrial plants, conventional modelling techniques and optimization tools are readily applicable for simulation and optimization of appropriate plant flowsheets for identifying the optimal design and operating conditions.

However, when facing a large design problem, the level of computational difficulty in the modelling and optimization, and the resources required to build models and obtain engineering solutions increase considerably. For such large industrial plants, a common practice is to decompose the design problem into several smaller subsystems to solve independently or sequentially. This division simplifies the overall problem and decreases the computational time required to find solutions. However, there is no guarantee that resulting solutions will be globally optimal, and they will be locally optimal with respect to certain objective functions and design constraints. For a non-integrated and simplified design model such as this, the complex design interactions related to plant retrofit and process changes will not be systematically investigated.

When a very detailed rigorous modelling and design framework of the plant is implemented, simulation and optimization of such models will typically require significant computational resources. Although various deterministic and stochastic optimization techniques have been successfully applied to a wide range of industrial applications (Smith, 2013 and Grossmann, 2013), it is often difficult to obtain optimal solutions with the available computational time and resources. Also, in-depth evaluation of economic trade-offs existing in the plant may not be envisaged when various design issues are interconnected and/or when different performance criteria are evaluated simultaneously. In particular, when the design problem is to retrofit the existing plant, it is more difficult to provide engineering solutions in practice due to inflexibilities associated with existing equipment size and process flowsheet configuration.

Heuristic rules or trial-and-error methods to screen retrofit alternatives or design options are often used in industries. Although such heuristic-based design procedures allow considerable engineers' flexibilities in building process models or carrying out economic trade-offs, these methods heavily relied on engineers' experience and/or insights for retrofit problems and, consequently, are limited for considering only manageable number of variables for the retrofit as well as for analyzing the process performance with simplified unit or process models rather than rigorous ones.

In order to systematically consider a large number of variables simultaneously and to accommodate impacts of structural changes in the configuration, optimization-based approaches had been widely adopted in the retrofit study. For example, Jackson and Grossmann (2002) developed a process synthesis framework which can model process flowhsheets and identify the optimal modifications in the flowsheet with the aid of mathematical optimization techniques.

On the other hand, attempts were also made to provide a systematic and robust basis for the judgement of economic and/or sustainable performance related to changes in operating conditions or structural modifications, and such indicator-based design method through a sensitivity analysis had been applied to the retrofit of chemical processes in generating and evaluating alternatives (Carvalho et al., 2008). To determine retrofit options for batch plant, Simon et al. (2008) proposed a design methodology using heuristics which is coupled with indicators of processing performance for three different levels, namely, plant, process and unit operation level.

For the retrofit study of continuous processes, automated design approach using optimization techniques had been widely used, for example, revamping study of crude oil distillation unit by Gadalla et al. (2013) in which attempts to provide retrofit solutions were made considering process changes and structural modifications together through rigorous simulation and optimization framework. Another example of using automated retrofit design based on stochastic optimizers is the retrofit of ethylene plant studied by Tahouni et al. (2013), in which determination of operating conditions for distillation columns was simultaneously made with refrigeration cycles and heat recovery systems for retrofit scenario. Deterministic optimization approach for finding retrofit options was also investigated by Chen et al. (2013) in which steam power plants were mathematically modelled and optimized for improving energy efficiency in steam and power production. Benefit of using optimization techniques in retrofit design was also demonstrated by the study of Krajnc and Glavic (2009) on finding optimal strategies for the co-production of bioethanol and sugar, which was systematically carried out with superstructure approach. On the other hand, engineering decision and analysis using process models and simulators had also played a part in the area of retrofit design. Polley et al. (2013) applied thermo-hydraulic simulation in the evaluation and analysis of heat exchanger network subject to fouling, with which assessing impacts related to structural changes and energy efficiency was made.

From recent studies on the retrofit of industrial processes, most of design methodologies developed were very specific to the industrial applications, for example, the retrofit of heat exchanger networks using intensified heat transfer equipment (Wang and Smith, 2013), the retrofit of cooling water systems using the reuse of cooling water (Reddy et al., 2013) and the retrofit of energy systems of a chemical plant using pinch technology (Feng and Liang, 2013). Conceptual understanding and design strategies obtained from these retrofit studies are very meaningful, but their generic application to other technologies or industries was inherently limited.

In order to overcome these shortcomings in the retrofit design of chemical processes, Hernández-Enríquez et al. (2011) proposed a generic strategy of Retrofit Design Approach (RDA) in which a process simulator, with the aid of Response Surface Methodology (RSM), is used to build a reduced model based on selected promising design variables, evaluate the impact of process changes, and find the optimal performance and operating conditions of the system. This design method allows the systematic assessment of various design scenarios for retrofit and provides a series of design options to be considered while using only modest computational effort.

Their design method investigated the optimization of operating conditions for the process and possible process modifications. However, potential modifications were considered mainly for particular pieces of equipment in isolation, for example the introduction of additional units in parallel with existing ones or changing the feeding stages of distillation columns. In retrofit scenarios, there are various design options including structural changes which modify the site-wide configuration of the plant. For example, distillation columns sequencing for multi-component separations can be reconfigured in a retrofit study by considering different order of separation, complex column arrangements, heat-integrated upgrading options (e.g. heat pumping), change of heat recovery between columns, etc. Such consideration were not fully explored.

Another important consideration to be made in the retrofit is to revamp the heat recovery systems. The reconfiguration of heat recovery systems is not only resulted from the energy saving study, but also changes made in the operating conditions due to retrofit. If operating conditions for a particular unit operation were changed during retrofit or debottlenecking of the plant, changes in heat recovery systems are, consequently, required. Therefore, the design methodology considering retrofit should reflect the plant-wide impact of re-configuration and/or changes of operating conditions in the design of energy systems.

In order to overcome drawbacks of Hernández-Enríquez et al. (2011)'s methodology, the methodology discussed in this paper allows systematic plant-wide investigation of energy usage and structural changes of the heat recovery network in a plant. Hence, the motivation for this work is to provide a conceptual design procedure which enables investigation of various design issues in a holistic manner in order to improve the performance of existing plants in the context of process retrofit.

The description of the design methodology employed in this study will be detailed in the next section, which is then followed by the case study for illustrating how the proposed design method is applied in practice as well as demonstrating how the plant-wide impact associated with changes made in the retrofit is considered. For the case study, the process is described giving process data, product specifications and economic parameters. This is followed by subsections describing the simulation model and retrofit objectives. Finally the proposed Retrofit Design Approach is then applied and results are presented and discussed.

DESIGN METHODOLOGY

To determine the most appropriate energy-efficient retrofit options, the design method proposed in this work employs a sequential approach proposed by Hernández-Enríquez et al. (2011) which consists of three stages, namely, diagnosis stage, evaluation stage and optimization stage. This methodology is able to handle retrofit problems, yielding investment portfolios with techno-economic improvements within a reasonable period of time. In addition, it provides insights into the most important factors affecting plant improvements and it also has flexibility with respect to the modelling tools integrated together.

The methodology proposed to address plant retrofit design is extended to four stages by adding design of energy recovery systems additionally, which allows achieving energy targets through the design of heat exchanger network. Also, it should be noted that methodological analysis and design procedures for the first two stages of diagnosis and evaluation now include the application of plant-wide process integration techniques, for example, energy-saving targeting using pinch technology, and their impacts on the process performance as well as profit. Through embedding systematic energy minimization within the retrofitting method, process improvements leading to profit increases are now achieved through not only stand-alone process upgrading but also changes in the design and/or operating conditions in an integrated manner. Once the process conditions have been fixed from these first three stages of analysis, the HEN retrofit may be implemented in the fourth stage. The design methodology proposed is summarized as:

Stage 1: Diagnosis

Any promising operating variables to be changed or options of structural modifications potentially leading to a cost-effective improvement in the process are first identified. This is done through monitoring one or more techno-economic key performance indices (KPI), which are specified based on the user's preference. In the current work, net profit and energy requirement were chosen to measure the degree of process improvements. Profit in this study is defined using the revenue from the sales of products and co-products minus the operating costs, including the raw material and the energy costs.

The design options leading to profit increase and/or energy savings can be screened with designers' experience and heuristics. However, it would be more effective for this purpose if a systematic approach could be employed. A wide range of process integration techniques are available for screening a large number of design options considered for retrofit study, for example, changing operating conditions, modifications of configuration and/or structure of the process, etc. A few selected process integration methods are listed in below, which can systematically evaluate different design options in the context of system-wide implication and economics, and provide the most promising options or settings of variables for the retrofit, subject to objectives and constraints.

- Composite hot stream and composite cold stream for identifying maximum potentials of energy recovery for the plant and providing design guidelines for the heat recovery systems (Smith, 2005 and Kemp, 2007).

- Design of heat exchanger networks for achieving the minimum energy requirement (i.e. widely referred as energy target) or improving energy recovery in the network configuration of heat exchanges (Linnhoff and Hindmarsh, 1983 and Ahmad et al., 2012).
- Total site targeting for assessing site-wide steam recovery and providing the design basis for site utility systems (Dhole and Linnhoff, 1993 and Klemes et al., 2010).
- Distillation systems design and optimization to consider complex columns arrangement and re-configuration of the sequence (Shah and Kokossis, 2002 and Jain et al., 2012).
- Reactor systems design and optimization to find optimal configuration and operating conditions for the reactor (Kokossis and Floudas, 1990 and Kokossis and Floudas, 1994).

A sensitivity analysis is performed through perturbations of the variables being studied and assessing their impacts on the profit and energy requirement. When the change of variables mainly affect energy consumption of the process, KPI for process integration methods in the sensitivity analysis is set to be energy requirement. For the retrofit options with which the consideration of capital cost is important, net profit is a more appropriate KPI for applying process integration methods.

At diagnosis stage, it would be ideal to consider as many process integration methods as possible. However, such rigorous application of wide range of design methods is not realistic in practice. Therefore, engineering judgement should be used when selecting the most appropriate process integration methods to be considered. For example, the case study considered in this paper is a fractionating plant, hence, three methods, namely, (i) composite hot stream and composite cold stream, (ii) design of heat exchanger networks and (iii) distillation systems design and optimization were considered.

The sensitivity analysis is carried out at this stage in order to gain a broad knowledge of the process, and not to select the most important factors. Therefore it differs from one-to-one analysis, which is not suitable for obtaining screening factors (Myers et al., 2004). The necessary criteria required to consider a variable as a promising one should be based on the order of magnitude of the change in response (i.e. profit and energy target) of the plant studied. In order to effectively identify promising improvements, these criteria are set by the user. Specifically, in the case study considered here improvements are considered promising if:

- The perturbation of that variable yields an increase in profit of at least 5% compared with the base case, without reducing the energy requirements.
- The perturbation of that variable achieves a combined increase of profits by at least 1% and reduction of energy requirements by at least 3% compared with the base case, and
- The perturbation of that variable does not affect the profit but yields a minimum 5% reduction of energy requirements compared with the base case.

From this sensitivity analysis the variables that do not show sufficient improvement (according to the above criteria) are removed and the remaining promising variables and options are considered in the following stage. As a consequence this first stage defines the initial size of the problem (i.e. the number of factors for the study), the ranges (or levels) over which these can be varied and any engineering constraints or practical restrictions to be considered in the retrofit.

Stage 2: Evaluation

The promising options found in the diagnosis stage are examined through a screening DoE (Design of Experiment), which is applied to identify the most important factors from the set of promising variables. Following this identification of important factors, a reduced model is obtained by fitting the process response behaviour. In most cases, additional simulations are required to accurately fit the surface model needed and to apply the RSM. The evaluation stage can be split in to two sections:

- *Preliminary screening* in which a Fractional Factorial Design (FFD) is carried out at "*n*" levels using "*k*" factors, where the number of levels "*n*" is chosen by the user and the "*k*" factors are the promising variables identified in the diagnosis stage. The first FFD screening identifies a set of process simulations which evaluate their impacts on the objective response. Analysis of variance (ANOVA) is subsequently applied to these responses from simulation which allows identification of the most important factors and fitting the surface model. RSM is performed in the second step of evaluation, based on the most important factors selected. The results are used for fitting, for example, with the aid of linear least squares method, which results in a reduced model. The reduced model obtained can reproduce data in the range studied, of which the accuracy is measured with Root Mean Square Error (RSME) and residual plots.

Stage 3: Process Optimization

A reduced model generated from the evaluation stage is now optimized to obtain the optimal value of the objective function and the optimal values of variables in the ranges studied. Various optimization solvers can be applied according to the characteristics of the reduced model and the level of nonlinearity in the model. Computational difficulty associated with getting optimal solutions is not a major issue for this case, because the resulting reduced model is based on a set of far simpler equations than a conventional optimization model which consists of various energy and mass balances formulated with highly nonlinear equations. However, it should be noted that computational efforts to obtain optimal solutions from reduced models are needed when a large number of variables should be dealt with or a reduced model involves severe nonlinearity.

Stage 4: Design of Heat Recovery Systems

The optimization of the process in the previous stage often results in changes of operating conditions or configuration in the process. Hence, the existing heat recovery systems should be reconfigured for accommodating such changes in the plant, or new possibilities to increase heat recovery are arisen with new operating conditions. Although it would be preferable to consider process and HEN optimization simultaneously (Smith et al., 2010), in this approach we adopt sequential optimization of the process and the HEN. Network pinch method (Asante and Zhu, 1996) is applied to identify promising structural changes to the network, including re-piping, re-sequencing, stream-splitting, and adding or deleting heat exchangers, through either by manual inspection or automated design methods. Several topologies can be obtained, which have different implications on capital investment for the retrofit and potential energy savings. Fig. 1 shows schematically the overall procedure of design method proposed in this study.

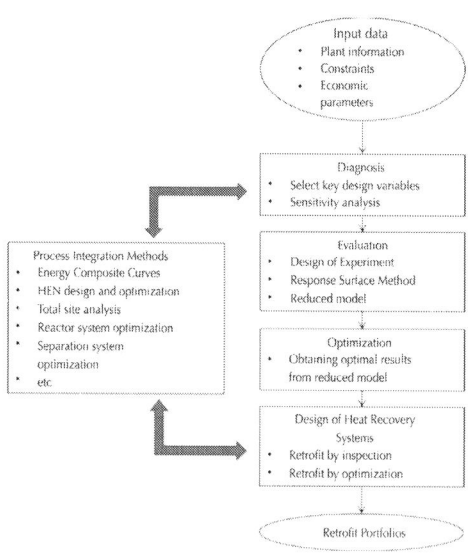

Figure 1: Overall procedure for Retrofit Design Approach.

CASE STUDY: PROCESS DESCRIPTION

To demonstrate the proposed approach it has been applied to a hydrocarbon fractionation (HCF) unit. This plant is presented in Fig. 2 and its main purpose is to separate two hydrocarbon rich feed streams into the products. This HCF plant is designed to process 105,000 barrels per day.

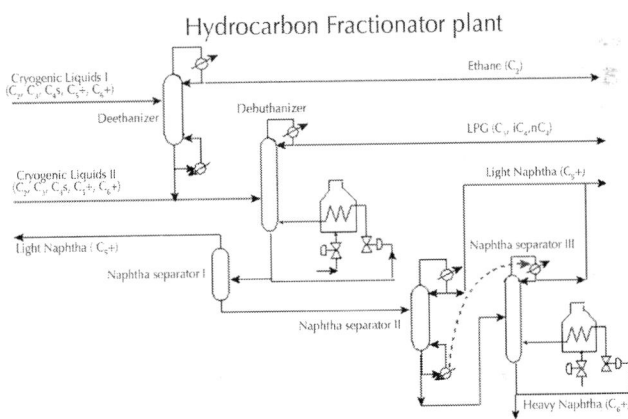

Figure 2: A hydrocarbon fractionation (HCF) process.

Two de-ethanizer columns set in parallel process feed supplied from cryogenic and liquid sweetener units, in Fig. 2 are simplified as "Deethanizer". Each column has 20 stages and in both cases the inlet is fed at the 6th stage of the column. The bottom products from the de-ethanizer columns and the C_{3+} feed from upstream process are mixed and fed to the 28th stage of the de-butanizer. Heat required to drive the separation is supplied to the bottom reboiler through a direct fire heater (i.e. gas furnace). The vapour product, which is rich in propane and butanes, is partially condensed by six cooling water condensers and the distillate is sent as feed to the de-propanizer column. The bottom products of the de-butanizer consist mainly of pentanes and heavier compounds (C_{6+}).

De-propanizer column is seldom used to produce coolant propane and therefore it is not shown in Fig. 2. However, to avoid operating difficulties and delays during the start-up of this column, it is preferred to keep its reboilers operating with LPS at low capacity and hence it is considered in the HEN analysis. The bottom products of the de-butanizer are sent to a series of naphtha separators. The first separator (naphtha separator I in Fig. 2) is a tank performing an initial split of light naphtha (C_{5+}) and heavy naphtha (C_{6+}). The second separator (naphtha separator II in Fig. 2) is a distillation column to separate and remove the light hydrocarbons, while the third one (naphtha separator III in Fig. 2) carries out the final reformation of naphtha in the distillation column.

Table 1 presents the feed flowrate, composition and operating condition. Product specifications are available in Table A1 in Appendix while 99% is for the required product recovery of ethane in the de-ethanizer, 98% for C3 product in the de-propanizer, C4 product in the de-butanizer and C5 product in the 1st naptha column, respectively, and 99% for C_{6+} product in the 2nd naptha column.

Table 1: Feed streams and LPG recovery

Feeds	Feed to de-ethanizers (cryogenic liquids I)	Feed to de-butanizer (cryogenic liquids II)
Composition		
C_1	0.0157	0.0001
C_2	0.3709	0.0098
C_3	0.2704	0.5928
nC_4	0.1168	0.1983
iC_4	0.0532	0.0822
nC_5	0.0504	0.0512
iC_5	0.0422	0.0429
C_{6+}	0.0800	0.0223
Flowrate, kg/s	45.36	13.42
Temperature, °C	40.8	30
Pressure, kPa	2255.5	1500.4

The objective function considers both the process and the HEN and is defined based on maximization of the annualized MPCA. In this study the annualized capital cost is calculated based on an interest rate of 12% and an expected 10 y project life. Also, the annual number of working days for the plant is considered to be 350 days per year, with a 30 day period of maintenance every two years. The available utilities and their costs are available in Table A2, and unit prices of the raw materials and products are available inTable A3 in Appendix. The capital costs of new units were estimated using Eq. (A6) with installation costs assumed to be 50% of the equipment costs (i.e. cost of purchasing equipment plus piping costs). All the capital cost information is based on the equations and correlations of Timmerhaus and Peters (2003).

CASE STUDY: APPLICATION OF THE PROPOSED RETROFIT DESIGN APPROACH

Diagnosis Stage: Selection of Key Design Variables

The independent variables in the plant comprised the initial set of variables to explore and the impact of these variables was assessed through the sensitivity analysis described in the previous section. Two responses were selected for study: the MPCA and the energy requirement. The criteria defined in the diagnosis stage were applied to select the promising variables.

The plant was simulated using AspenPlus® simulator 2006.5 with the sequential modular mode (SMS) setting and with the Peng–Robinson method for the calculation of thermodynamic properties. All the flowsheets were simulated using standard modules available in the AspenPlus® library.

The process stream data is extracted from the Aspen Plus simulator report sheets and this data is typed into SPRINT® (version 2.4.001) as shown in Fig. 3. When minimum approach temperature (T_{min}) of 10 °C is taken for the targeting of maximum energy recovery for the process considered, the minimum hot utility requirement is 42.570 MW and the minimum cold utility requirement is 33.489 MW.

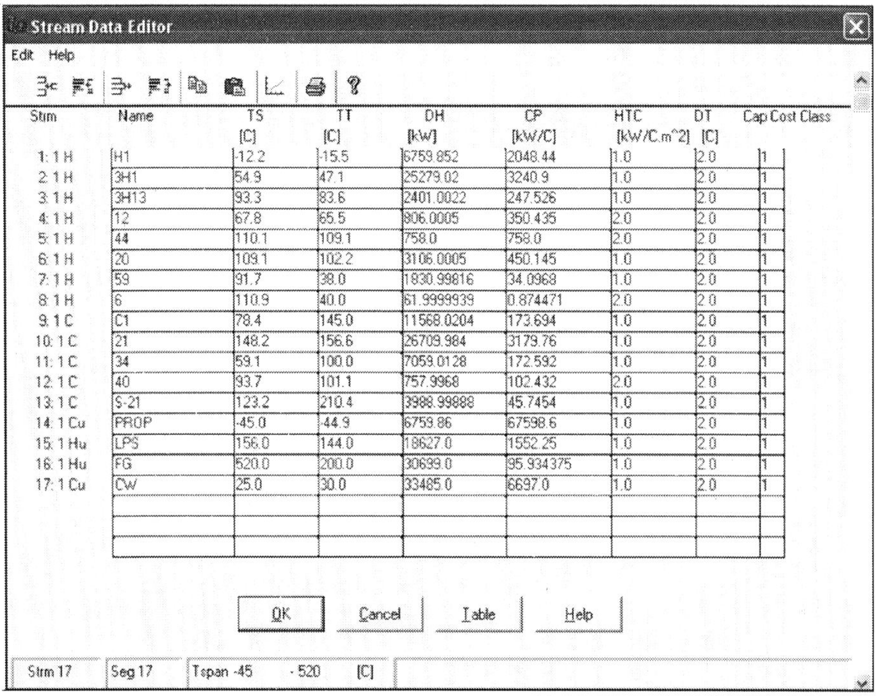

Stm	Name	TS [C]	TT [C]	DH [kW]	CP [kW/C]	HTC [kW/C.m^2]	DT [C]	Cap Cost Class
1: 1 H	H1	-12.2	-15.5	6759.852	2048.44	1.0	2.0	1
2: 1 H	3H1	54.9	47.1	25279.02	3240.9	1.0	2.0	1
3: 1 H	3H13	93.3	83.6	2401.0022	247.526	1.0	2.0	1
4: 1 H	12	67.8	65.5	806.0005	350.435	2.0	2.0	1
5: 1 H	44	110.1	109.1	758.0	758.0	2.0	2.0	1
6: 1 H	20	109.1	102.2	3106.0005	450.145	1.0	2.0	1
7: 1 H	59	91.7	38.0	1830.99816	34.0968	1.0	2.0	1
8: 1 H	6	110.9	40.0	61.9999939	0.874471	2.0	2.0	1
9: 1 C	C1	78.4	145.0	11568.0204	173.694	1.0	2.0	1
10: 1 C	21	148.2	156.6	26709.984	3179.76	1.0	2.0	1
11: 1 C	34	59.1	100.0	7059.0128	172.592	1.0	2.0	1
12: 1 C	40	93.7	101.1	757.9968	102.432	2.0	2.0	1
13: 1 C	S-21	123.2	210.4	3988.99888	45.7454	1.0	2.0	1
14: 1 Cu	PROP	-45.0	-44.9	6759.86	67598.6	1.0	2.0	1
15: 1 Hu	LPS	156.0	144.0	18627.0	1552.25	1.0	2.0	1
16: 1 Hu	FG	520.0	200.0	30699.0	95.934375	1.0	2.0	1
17: 1 Cu	CW	25.0	30.0	33485.0	6697.0	1.0	2.0	1

Figure 3: Stream data.

Column Operating Pressures

For the plant considered, column operating pressures were considered as continuous variables determine if a change in operating pressures can lead energy savings or profit increase. Focus was made to improve energy efficiency rather than increasing the profit for this case, because the net profit of the plant, except energy cost, is not heavily influenced by the change of column operating pressures.

Plus–minus principle (Smith, 2005) was considered by manipulating operating conditions of stream in order to improve heat recovery without compromising in the profit and product qualities. All the streams were screened with the aid of plus–minus principle, and sensitivity analysis was made to evaluate impacts associated with change in column operating pressures. Three promising variables were selected for further investigation in the retrofit study on the basis that improvement in energy target (i.e. more than 3% of reduction in either hot utility consumption or cold utility consumption) as shown in Table 2.

Table 2: Promising continuous factors for the case study

Unit	Variable	Variable range	Value leading to max. energy saving	Variation in MPCA, %	LPG recovery, %	Hot utility target, %	Cold utility target, %
De-ethanizer column	Pressure in top, kPa	784.5–1784.8	784.5	+1.6	+0.01	−3	−3
De-butanizer column	Pressure in top, kPa	833.6–1569.1	1569.1	+1.8	+0.01	−2	−6
Naphtha separator II	Stage 1 Pressure, kPa	343.2–411.9	343.2	+0.2	0	−4	−6

+, increased; −, decreased.

Column Feed Stage

The location of feed for all columns in the system were explored, and their sensitivities were tested by considering the same feeding stage, feeding to 3 stages above and feeding to 3 stages below the current position. Given the sizes of the four columns this variation was sufficient for evaluating its economic impact. The results of sensitivity analysis for a de-ethanizer column are shown in Fig. 4 which presents composition profiles and with which internal re-mixing effects can be examined. The composition profiles do not show any evidence of significant improvement in the de-ethanizer column, therefore it was concluded that the change of feed stage position for this column is not considered as a promising variable. In a similar manner, de-butanizer and two naphtha columns were assessed and no considerable potential for improvements by changing the feed stage position for these columns was found. Therefore, change in feed stages for the columns was not considered in the retrofit study.

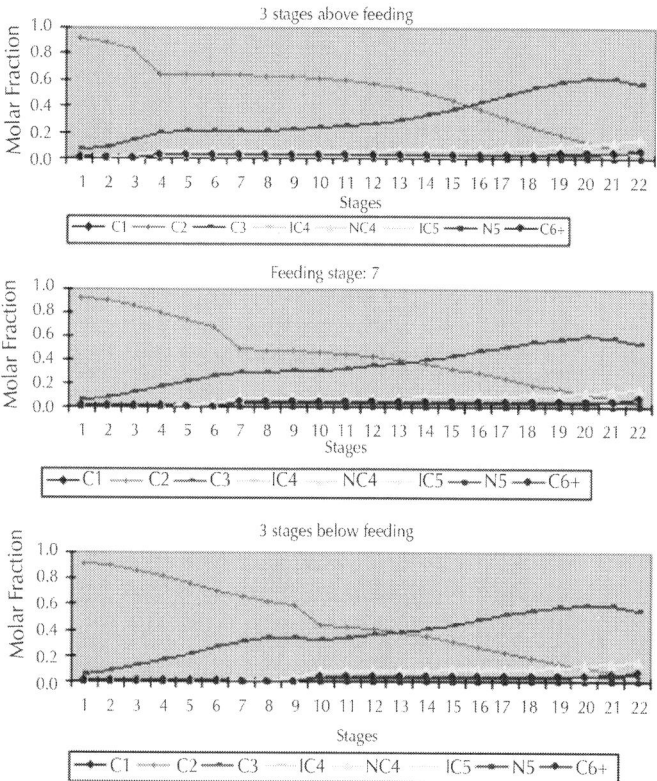

Figure 4: Composition profiles for a de-ethanizer column.

Thermal Conditions of Column Feed

The thermal condition of the feed to the column was considered. Current thermal condition of the feed to the de-ethanizer is saturated liquid. VF represents the degree of dryness in the feed (i.e. VF = 0 is saturated liquid, while VF = 1 is saturated vapour). Various values of VF were tested, and some of results simulated with different VF values are presented in Fig. 5, which indicated that increasing the vapour fraction in the feed decreases ethane fraction the column, reduces the purity of the top product as well as LPG recovery. Hence, it is concluded that changing inlet conditions for the de-ethanizer from saturated liquid to high VF value is not considered.

For the de-butanizer column, the value of VF in the base case is 0.01, and range between 0.25 and 1 was considered. No improvements were observed. Naphtha separator II column was currently operated with 0.03 of VF, and it was changed between 0.25 and 1. There was no evidence of improvement for the separation of light naphtha (iC_5 and nC_5) and heavy naphtha (C_{6+}) in this column. Thus, this variable was not considered in the evaluation stage.

For naphtha separator III column, the VF value was 0 initially, and changes between 0.25 and 1 were considered. No significant improvement exists for the separation of light and heavy naphtha either. For all the columns with respect to the inlet feed conditions (VF), the MPCA response did not show a significant improvement. Also, these changes of thermal feed conditions would require structural modification of the heat recovery systems. Consequently, this option was not considered in the following evaluation stage.

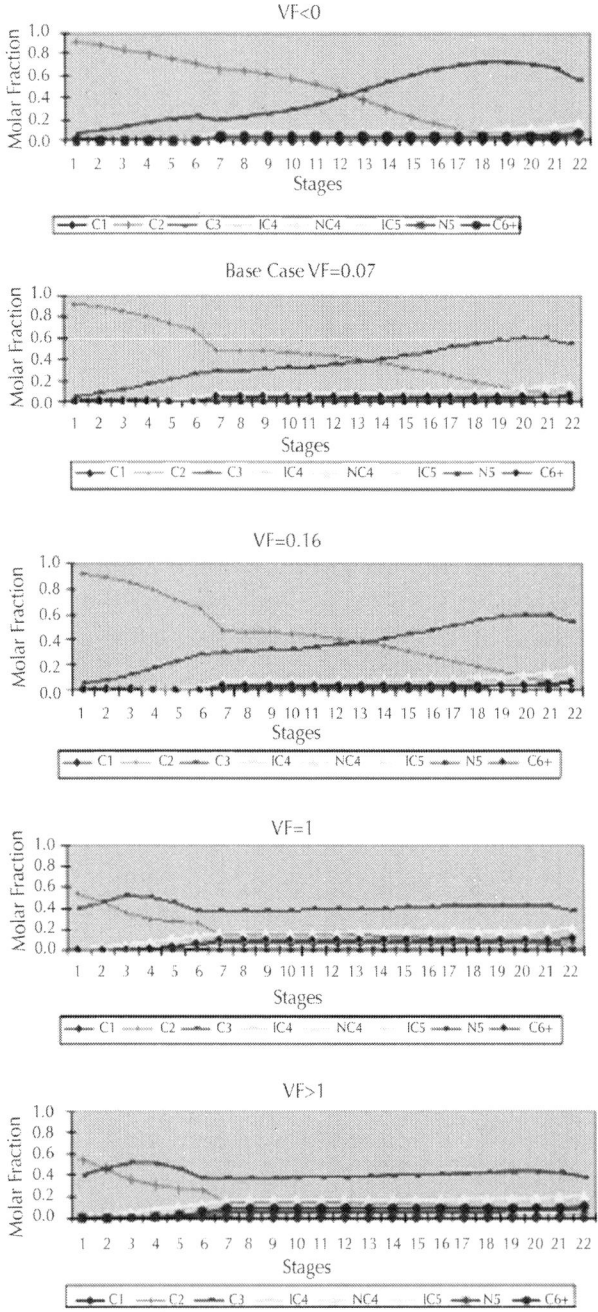

Figure 5: Effect of feed condition for the de-ethanizer column.

Column Stage and Efficiency

Two options were investigated here: an increase in the number of stages and upgrading of column internals to give higher efficiency. For this purpose the magnitude of the perturbations were chosen to be the addition of 5 stages and a stage efficiency increase of 20% for all the columns which is based on upgrading of efficiencies possible in the current market (Koch-Glitsch, 2013). If the efficiency has a considerable impact during the evaluation, the different number of trays should also be tested and the most appropriate number of stages selected. For the de-butanizer and in the naphtha columns, impacts from the increase of the number of stages and efficiencies were very low which were not considered further. The only promising options identified are an increase in the number of stages and the stage efficiency for the de-ethanizer column, which is summarized in Table 3.

Table 3: Effects of the number of stages and their efficiency

Description	Range	Best setting	MPCA, %	LPG recovery, %	Hot utility target, %	Cold utility target, %
Number of stages for the de-ethanizer column	20–25	25	+1.3	0	−9.3	−16
Stage efficiency of the de-ethanizer	58–78%	78%	+1.8	+0.01	−11.6	−20

+, increased; −, decreased.

Distillation Re-Sequencing

A series of feasible distillation arrangements were screened and simulated. Because of high purchase cost of new columns, the retrofit focused on arrangements that could re-use the existing columns in the plant. To make this possible, it was needed to take into consideration both the size of the columns in the plant and the possible distillation sequencing arrangements.

Since the production of propane is seldom required, options for re-sequencing or re-configuration are also not considered. Also, options for considering structural changes of three columns simultaneously were excluded due to practical issue during the implementation. Hence, the following the set of possible arrangements were considered:

- Re-sequencing option 1: a sloppy arrangement with two de-ethanizers and a de-butanizer.
- Re-sequencing option 2: a prefractionator arrangement between a de-ethanizer and a de-butanizer.
- Re-sequencing option 3: a prefractionator arrangement between a de-butanizer and a naphtha separator II.
- Re-sequencing option 4: a prefractionator arrangement between a naphtha separator II and a naphtha separator III.

Re-sequencing option 3 is not considered because the de-butanizer is currently integrated with the furnace, which makes difficult to employ a prefractionator arrangement with naphtha separator II. A prefactionator arrangement is typically favoured when intermediate products make up a large proportion of the feed (e.g. more than 50%). However, middle boiling point components (nC_5) account for only around 26% by mole fraction in the feed for naphtha separator II, hence re-sequencing option 4 is not considered further. Also, it should be noted that side-stream columns were not considered in this case study. This exclusion is based on studies (Smith and Linnhoff, 1988) such that side columns are favoured when intermediate product is dominant, and either the lightest or the heaviest product is present in small quantities, typically, less than 5%.

After screening of available options, two promising arrangements were selected for further analysis: re-sequencing option 1 (Fig. 6) and re-sequencing option 2 (Fig. 7).

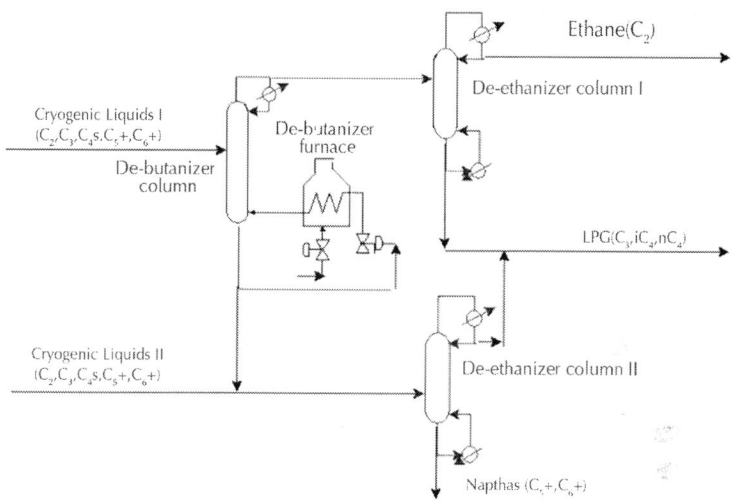

Figure 6: A sloppy arrangement with two de-ethanizer and de-butanizer columns.

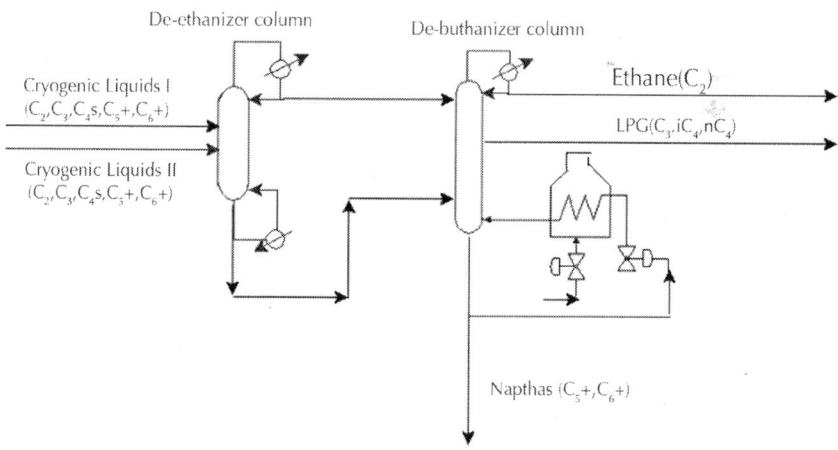

Figuer 7: The prefractionator arrangement between one de-ethanizer and de-butanizer columns.

It is important to mention that in the re-sequencing option 1 (Fig. 6) the de-butanizer column was installed in first place to be able to operate the two following de-ethanizer columns at the same or similar operating pressure and to extract the middle product as a side stream.

On the contrary, for the re-sequencing option 2 (Fig. 7), the de-butanizer column was set at the end because of its size, which is larger than one of the de-ethanizer columns used. Therefore its capacity is large enough to handle an additional intermediate products stream (LPG).

Table 4 presents a summary of the results for two re-sequencing options considered in details. For both options, the location of feed stages was varied for seeking any further improvement, but no significant improvements were found. For option 1, the energy consumption is higher than that of the base case due to the operation of two de-ethanizer columns, instead of one and because one of the columns is processing more heavy components than in the base case. Despite this increased duty and a minor increase in the LPG recovery the MPCA is reduced 10.6%. Additionally, the presence of 2.2% (volume) butanes in the C_{5+} product is the major disadvantage because the maximum allowed for butanes in the product is 2% (volume).

Table 4: Simulation results for distillation re-sequencing options

Distillation arrangement	Butane content in C_{5+} product[a]	MPCA, %	LPG recovery, %	Variation duty, %
Base case	None	0%	0%	0%
Re-sequencing option 1	2.2% (vol)	−10.6	+0.07%	+7.6%
Re-sequencing option 2	4.5% (vol)	+27%	−1%	−43.6%

[a]Maximum 2% (vol) allowed.

For option 2, the location of feed and product streams in de-butanizer is made as feed stream coming from the bottom of de-ethanizer at stage 30, feed stream coming from the top of de-ethanizer at stage 5, C_3 product extraction at stage 16 and C_4s product extraction at stage 34. The energy consumption is significantly reduced when compared with the base case, due to the considerable reduction of re-mixing effects inside the de-butanizer column. From Table 4, significant reduction in the heat duty and increases in the MPCA are obtained, despite a slight decrease in the LPG recovery. However, option 2 does not meet the product specification in terms of amount of butanes component in the C_{5+} product, although various attempts by changing the flowrate and operating conditions were made to reduce butanes in the C_{5+} product stream.

Although the prefractionator arrangement can be energy-efficient it has not been considered further here due to off-specification of one of the final products. If the C_{5+} product sales are relatively small and energy cost is high, this prefractionator arrangement could be considered as a promising option at the expense of the penalty associated with off-specification.

Following this pre-screening of various options and variables through sensitivity analysis, promising factors which increased considerably profits and/or decreased energy requirements were selected, as summarized in Table 5.

Table 5: Variables for the 5 factors at 2 levels used in Full FD

Factor	Description	Units	Level (−1)	Level (+1)
Continuous variables (operational changes)				
X_1	Column pressure of the de-ethanizer	kPa	784.5	1784.8
X_2	Column pressure of the de-butanizer	kPa	833.6	1569.1
X_3	Column pressure of the naphtha separator II	kPa	343.2	411.9
Discrete variables (structural changes)				
X_4	Increase the number of stages for the de-ethanizer	Number	20	30
X_5	Increase the stage efficiency for the de-ethanizer	%	58	78

Evaluation Stage

Preliminary Screening

The promising variables listed in Table 5 were assessed to understand their detailed impact on the response MCPA, which intrinsically considers the energy requirements. The initial size of the problem is five factors, with at least two possible levels considered for each factor. No geometrical restrictions on the outputs for the searching space (i.e. geometrical form) were identified in the diagnosis stage. To identify the most important factors, a screening full factorial DOE was applied using Matlab® based on the five factors (k) and two levels with a maximum number of runs ($2k$) using the function "ff2n". The generators were: 'X_1', 'X_2', 'X_3', 'X_4', 'X_5', 'X_1X_2', 'X_1X_3', 'X_1X_4', 'X_1X_5', 'X_2X_3', 'X_2X_4', 'X_2X_5', 'X_3X_4', 'X_3X_5', and 'X_4X_5'. The resolution level for this design was "V" which provides a good balance between the number of runs and the confounding level. The number of simulations implemented by the generated design was thirty-two. The corresponding natural variables (real operational values for each factor at the two levels), are presented in Table 5. Interactions between three or more factors were considered to have a smaller effect on the MPCA response and therefore these were not taken into consideration.

The analysis of variance (ANOVA) for the 32 simulation responses was carried out using the statistic toolbox of Matlab® 7.0.1 with a significance level set at 99.5%. Table 6 presents the results for the main factors, from which the last column shows the p-values for each factor. Factors X_1, X_2 and X_5 were observed to have p-value less than 0.005 and hence these are identified as the most important.

Table 6: ANOVA results for main factors

Factor	Sum of squares	F test value	p-Value	Effect of the factor

X_1	1.05254	95.59	0	0.364
X_2	1.06116	96.38	0	−0.363
X_3	0.05351	4.86	0.0365	0.198
X_4	0.10322	9.37	0.0051	0.104
X_5	0.31508	28.62	0	0.082
Error	0.28628			
Total	2.89115			

The ANOVA results for all the second-order interactions of the 5 factors gave p-values larger than 0.005 and hence these interactions are not significantly important. From Table 6, the p-values of X_1, X_2 and X_5 were set at zero value. Thus in order to properly rank them in their order of importance and to verify the results provided by the ANOVA, the effect of each of all the factors as a main or second-order interaction was estimated as:

Effect of factor(i) = $\bar{Y}(+) - \bar{Y}(-)$ (1)

with $\bar{Y}(+)$ denoting the average of all response values (MCPA) for which factor i is in the "+1" level, and $\bar{Y}(-)$ denoting the average of all response values (MCPA) for which factor i is in the "−1" level of the DOE.

Fig. 8 plots the effect of all the main factors and second-order interactions on MPCA, which confirms the results of the ANOVA as factors X_2 (column pressure of the de-butanizer), X_1 (column pressure of the de-ethanizer), and X_5 (efficiency of stages in the de-ethanizer column) fall outside the specified limit indicated by the dashed line (i.e. ±0.1 for the effect on MPCA).

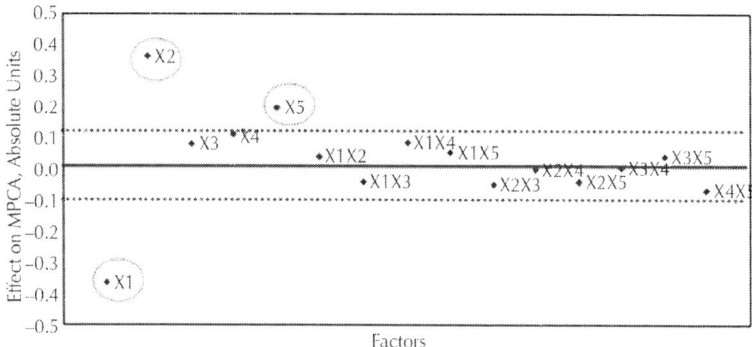

Figure 8: Effect of factors on MPCA response.

Application of RSM

A Central Composite Design (CCD) was built for the three most important factors with additional points placed at the α values of 1.316 as α = ± (3)$^{1/4}$ = ±1.316. The consideration made in the RSM for the CCD limits were:

Factors X_1 and X_2: The maximum limit was set as the coded level +1.316, and the minimum limit was set as the coded level −1.316 (operational limits). The rest of levels in the design were kept as in the screening DOE.

Factor X_5: The maximum value offered in the market for this class of column (85%) was set as the coded level of +1.316. For the coded level of −1.316, the current efficiency was taken (58%). The complementary levels in the design were ranged between these two limits.

Fifteen simulations were run following this CCD design. The MPCA responses were obtained and a linear least squares (LLS) method was used in Matlab® to fit the corresponding model. The best fitting model with a RMSE of 0.93 £/y (6.64% of the base case MPCA) is shown below where MPCA is given as a function of factors X_2, X_1 and X_5 (i.e. coded values between 0 and 1.316):

$$MPCA = 0.93 + 0.05X_1 - 0.01X_2 - 0.07X_5 - 0.03X_1^2 - 0.04X_2^2 + 0.03X_5^2 \quad (2)$$

The RMSE indicated an acceptable level of accuracy for the MPCA response and this is verified by the residual plot in Fig. 9. Hence, this response surface model is suitable to predict the simulation response MPCA.

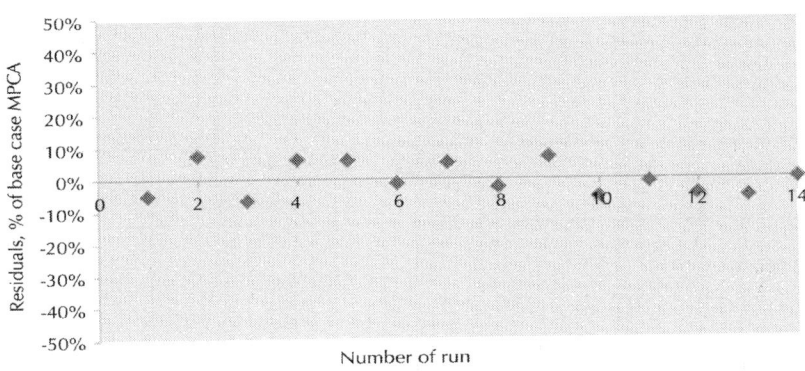

Figure 9: Plot of residuals for the best fit model in the HCF plant.

Optimization Stage

To avoid operational problems (e.g. surges in the ethane compressors and column flooding) the optimization was carried out within normal operating limits (i.e. +1 and −1 levels) for the factors X_1 and X_2. The factor X_5 was also set in the +1 and −1 range to take into account the difficulties associated with achieving the maximum/minimum efficiencies offered by the suppliers under the current conditions. It was possible to maximize the best fit model using NLP optimization to varying the three variables simultaneously using the Excel® solver. To ensure reliability and robustness of the solutions, optimization was initiated from different starting points which yielded different maximal points. The final maximum optimum conditions achieved with the coded and corresponding natural variables are presented in Table 7. Additionally, a set of simulations were performed around this maximum obtained with the reduced model to be sure that the maximum found was a "true maximum". The percentage difference between the reduced model and the simulation at the optimal conditions is shown in the last column of Table 7.

Table 7: Coded and natural variables for the optimal results

Coded variables			Natural variables			Difference between a reduced model and simulation at optimal conditions
X_1	X_2	X_5	X_1 kg/cm²	X_2 kg/cm²	X_5 fraction	
−1	1	1	8	15	0.78	0.062%

The detailed conditions of the optimum case are presented in Table 8, including main product compositions, total hot and cold utility requirements and the relative change in LPG recovery and normalized MPCA (MCPA*). Fig. 10 depicts the flowsheet of the plant with base case and optimum case data. The optimum case has 4.7% higher than the base case in terms of MPCA* and this is mainly due to reduced hot utility consumption (−4.4%), reduced cold utility consumption (−1.8%) and increased LPG recovery (+1%).

Table 8: Optimal results vs. base case

Product streams	Ethane gas			LPG			C_{5+}		
Mole fraction	BC	OC	% diff.	BC	OC	% diff.	BC	OC	% diff.
C_1	0.0478	0.0543	13	0	0	0	0	0	0
C_2	0.9329	0.9386	0.61	0.0129	0.0083	−36	0	0	0
C_3	0.0193	0.0071	−63	0.6265	0.6313	0.8	0	0	0
nC_4	0	0	0	0.2484	0.2482	−0.1	0.0095	0.0090	−5
iC_4	0	0	0	0.1122	0.1122	0	0.0003	0.0003	0
nC_5	0	0	0	0	0	0	0.2736	0.2727	−3
iC_5	0	0	0	0	0	0	0.2042	0.2053	1
C_{6+}	0	0	0	0	0	0	0.5124	0.5127	0.1

Flow-rate, kg/s	10.23	10.23	0	29.15	29.15	0	18.89	18.89	0
T, °C	−15.4	−33.3	117	30.7	30.7	0	38	38	0
P, kPa	1422.0	784.5	−45	706.1	706.1	0	843.4	843.4	0
Total hot utility, kW							40.6	38.9	−4.4
Total cold utility, kW							39.0	38.3	−1.8
LPG recovery, %							90.5	91.5	+1
MPCA, absolute units							1	1.047	+4.7

BC, base case; OC, optimum case.

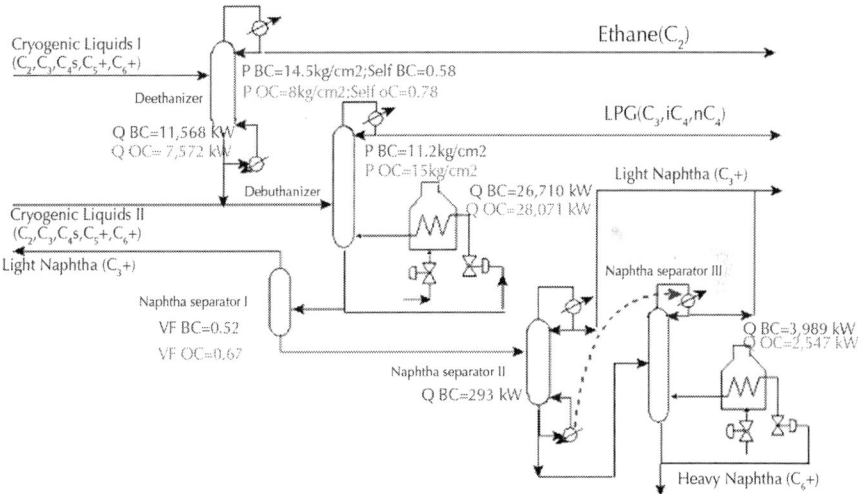

Figure 10: Optimal results vs. base case.

HEN Optimization

Since the de-propanizer column is rarely operated to produce propane, it was not taken into account in the previous sections. However, due to availability and the easiness of starting operations when propane is requested to be produced, the de-propanizer reboilers operate continuously using LPS at reduced capacity (affecting the MPCA) and hence they are considered in the HEN retrofit. Fig. 11 shows existing HEN configuration and the cross-pinch report from SPRINT® highlights inefficient heat recovery in the exchanger network. There were five exchangers transfer heat across the pinch in total of 7.371 MW, which are 3.121 MW for HX3, 2.421 MW for HX6, 0.701 MW for HX7, 0.035 MW for HX8, 0.989 MW for HX9 and 0.102 MW for HX11. Both retrofit by inspection and retrofit by automated design were explored with the objective to minimize the total cost.

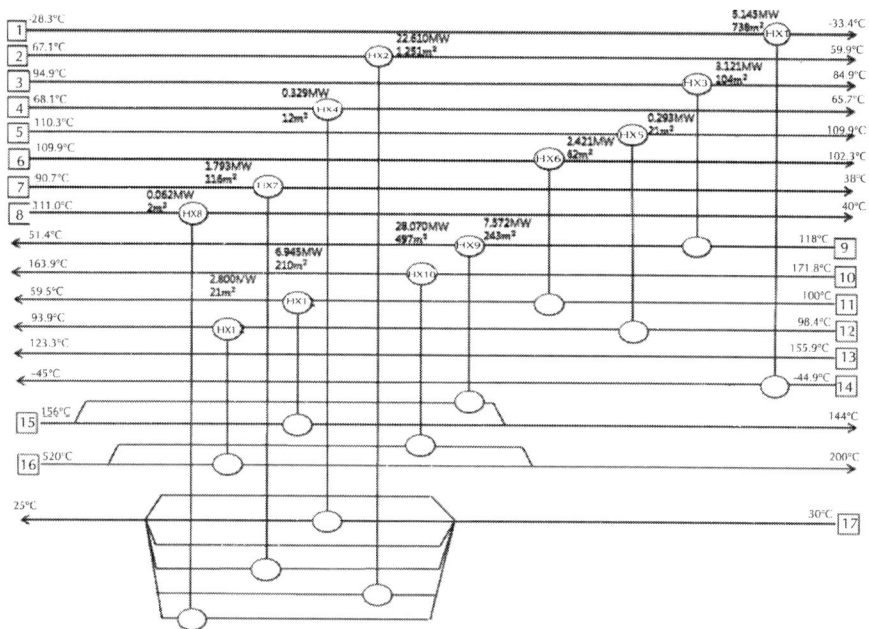

Figure 11: Existing HEN (heat exchanger network).

Exchangers 3, 6, 7, 8, 9 and 11 which transfer heat across the pinch were examined by inspection. The order of priority for removing existing cross-pinch exchangers was firstly re-sequencing, then re-piping, and finally adding a new match (i.e. new heat exchanger). The introduction of new utility paths and stream splitting were also explored and practical constraints were considered to avoid any expensive and non-practical options, including maximum number of re-sequencing = 2, maximum number of re-piping = 4, maximum number for splitting streams = 2, and maximum number of new match = 2.

Retrofit design by inspection was carried out looking for the highest cost-benefit balance of each modification. The changes were implemented one by one sequentially using the SPRINT® software. The two most promising retrofit options obtained by inspection involve re-piping and their topologies are shown in Fig. 12(a). Fig. 12(a) shows two retrofit modifications, including (1) Exchanger 3 is re-piped from a cooling water stream number 17 to pre-heat the bottoms of a de-ethanizer column (stream 9) and (2) Exchanger 6 re-piped from a cooling water stream 17 to pre-heat the bottoms of the de-propanizer column (stream 11).

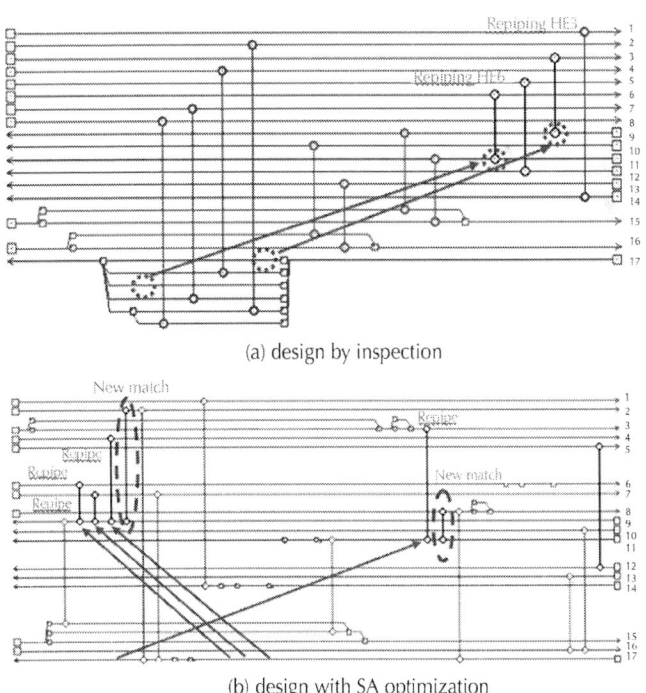

Figure 12: Modified HEN.

Retrofit through automated design was performed using the SPRINT® software with the simulated annealing (SA) parameters given in Table A4 and Table A5 in Appendix. The optimum solution obtained through SA-based optimization design is given in Fig. 12(b) and Table 9. The capital costs for the new matches were estimated by Eq. (A6) with installation costs assumed to be 50% of the equipment costs, which were estimated based on a correlation from Timmerhaus and Peters (2003). Re-piping and other costs were estimated to be 50% of the exchanger capital cost.

Table 9: Comparison of results for HEN retrofit design

Method	Current HEN without retrofit design	HEN retrofit design by SA	HEN retrofit design by inspection
New area, m²	0	693	152
Number of matches eliminated	0	0	0
Number of re-sequencings	0	0	0
Number of re-pipings	0	4	2
Number of new matches	0	2	0
Annualized capital cost, MM£/y	0.234	1.514	0.605
Energy savings, MM£/y	0.6	2.1	1.6
% improvement in MPCA	4.7	17.3	13.4
Payback period, y	0.438	0.731	0.377
Potential CO_2 tax reductions, MM£/y	0.4	1.4	1.1

Options presented in Table 9 provide a meaningful base for decision-making in capital investment because it summarizes the potential improvements, difficulties and payback period of each design proposed. Nevertheless, the final decision will depend on the budget and priorities of the final users.

Consideration of the Level of Capital Investment

A portfolio of HEN retrofit designs at various levels of capital investment were generated based on the modified process with optimal values for three most important factors. Two of these factors are pressures of the de-ethanizer and de-butanizer columns and they do not require a major change in the columns. The third factor is the efficiency of stages in the de-ethanizer and this modification requires a major change in the column internals. Hence, a decision should be made whether capital is invested for column modifications. This decision should be further linked with the retrofit design of HEN, as column modifications result in different stream conditions for heat integration study, leading to different optimal results in the HEN. Therefore, six different cases can be generated, which have different economic impacts and level of capital investment (Table 10).

Table 10: Feasible options for retrofit design

Cases	Case A	Case A1	Case A2	Case B	Case B1	Case B2
Column modification considered?	No	No	No	Yes	Yes	Yes
HEN retrofit	No	Retrofit design by inspection	Retrofit design by SA optimization	No	Retrofit design by inspection	Retrofit design by SA optimization
New Area, m²	0	152	693	0	161	713
Number of matches eliminated	0	0	0	0	0	0

Number of re-sequencings	0	0	0	0	0	0
Number of re-pipings	0	2	4	0	2	4
Number of new matches	0	0	2	0	0	2
ACC, MM£/y	0.234	0.605	1.514	0	0.371	1.281
Energy Reductions, MM£/y	0.6	1.6	2.1	0.5	1.5	2.0
% improvement MPCA	4.7	13.4	17.3	3.9	12.1	16.7
Payback period, y	0.438	0.377	0.731	0	0.256	0.641
Potential CO_2 tax, MM£/y	0.4	1.1	1.4	0.4	1.0	1.4

From Table 10, the HEN retrofit designs achieved in the Cases A1 and B1, and Cases A2 and B2, were very similar, except that for Cases B1 and B2 a larger additional heat exchange area is required than for Cases A1 and A2, respectively. Although the final decision may depend on non-technical factors, the information presented in the Table 10 provides a portfolio of feasible options based on the Retrofit Design Approach developed.

CONCLUSIONS

The Response Surface Methodology (RSM) based retrofit design framework originally proposed by Hernández-Enríquez et al. (2011) has been extended to address highly complex process retrofit design problems and to consider the design of heat recovery and its design. The new framework now considers structural changes modifying the site-wide configuration of the plant and now also considers the subsequent retrofit and integration of the integrated heat exchanger network. This framework can be applied to large-scale plants to identify promising and profitable retrofit design options. The final result integrates all the promising retrofit possibilities into a portfolio which can be used to make informed retrofit decisions.

The proposed Retrofit Design Approach has been successfully applied to an industrial case involving a hydrocarbon fractionator plant, and the approach has proven to be reliable and practical to achieve pseudo optimal solutions within a reasonable timescale. The results are based on the output of commercial simulators and statistical analysis, together with conceptual knowledge the process and process integration concepts. Applying the methodology to the hydrocarbon fractionator plant identified a portfolio of different retrofit options with varying levels of capital investment. All these options gave increased profits, including the option with no capital investment. However, the addition of increased capital investment allows for options giving bigger profits and reduced energy and utility requirements. As the proposed method in this paper involves a series of decomposed sub-problems and investigates important decision variables for each sub-problem, such application, especially for large-size retrofit problems, may not be computationally trivial and possibly requires users' good understanding on process integration methods. Therefore, further development is needed to systematically apply process integration methods in a more simplified and automated manner which less relies on users' judgement. Also future work could address the integrated design of both the process and the heat exchanger network simultaneously, rather dealing with two design problems in sequence.

ACKNOWLEDGMENTS

The second and third authors acknowledge that this research was supported by the International Research & Development Program of the National Research Foundation of Korea (NRF) funded by the Ministry of Science, ICT & Future Planning of Korea (No. 20110031290).

APPENDIX A. BASIS FOR ECONOMIC EVALUATION

For this study five types of profit used in Hernández-Enríquez et al. (2011) were adopted and its descriptions are given as:

$$NPr = SP + SCO - VCRM - VCE \tag{A1}$$

where NPr is the net profit [£/y], S_p is the profit from the sales of products [£/y], S_{CO} is the profit from the sales of co-products [£/y], VC_{RM} is the cost of raw materials [£/y] and VC_E is an Energy cost [£/y].

$$MaPr = NPrSC - NPrHBC \tag{A2}$$

where MaPr is the marginal profit [£/y], NPr_{SC} is the net profit of studied case [£/y] and NPr_{HBC} is the net profit of historical best case [£/y].

$$MaPr^* = \frac{MaPr}{MaPr_{BC}} \tag{A3}$$

where MaPr* is the marginal profit normalized [–], MaPr is the marginal profit of studied case [£/y] and $MaPr_{BC}$ is the marginal profit of base case [£/y].

$$MPCA = MaPr - ACC_i \tag{A4}$$

where MPCA is the marginal profit capital Affected [£/y], MaPr is the marginal profit [£/y] and ACC*i* is the annualized capital costs of "*i*" change suggested for retrofit [£/y].

$$MPCA^* = \frac{MPCA}{MPCA_{BC}} \tag{A5}$$

where MPCA* is the marginal profit capital affected normalized [–], MPCA is the marginal profit capital affected of studied case [£/y] and $MPCA_{BC}$ is the marginal profit capital affected of base case [£/y].

To consider the capital costs associated with the structural changes proposed in the retrofit, capital investment is annualized as annualized capital cost for new units ($ACC_{NewUnits}$):

$$\text{ACC}_{\text{NewUnits}} = \text{CC}_{\text{NewUnit}} \cdot \text{AF} \tag{A6}$$

where $\text{ACC}_{\text{NewUnit}}$ is the annualized capital cost of the new unit [MM£/y], $\text{CC}_{\text{NewUnit}}$ is the Capital Cost of the new unit (acquisition cost plus piping cost plus installation cost) [MM£] and AF is the Annualization Factor [y^{-1}].

Finally, the payback period and the reduction of carbon tax achieved with the proposed retrofit design options can be estimated. The payback period for general profit improvements is estimated as:

$$\text{PP} = \frac{\text{ACC} \cdot \text{PLP}}{(\text{PI})} \tag{A7}$$

where PP is the payback period [y], ACC is the annualized capital cost [MM£/y], PLP is the Project Life Period [y] and PI is the general profit improvements [MM£/y].

The payback period for energy improvements is calculated as:

$$\text{PP} = \frac{\text{ACC} \cdot \text{PLP}}{(\text{EI} + \text{CO}_2\text{T})} \tag{A8}$$

where PP is the payback period [y], ACC is the annualized capital cost [MM£/y], PLP is the Project Life Period [y], EI is the energy improvements [MM£/y] and CO_2T is the annualized benefit from reduction in CO_2 emission taxes [MM£/y].

Energy improvements (the reduction of energy costs) are directly quantified using the reduction of energy requirements multiplied by the utility unit cost.

The reduction of carbon tax is derived mainly from the energy savings in the process. CO_2 emissions are estimated based on the energy reduction using the EPA 42 factor for furnace combustion (US EPA, 2006), and the resulting CO_2 flowrate is multiplied by an estimated cost of CO_2, based on the Estimated Social Cost of Carbon (SCC) for 2005 of USD \$43/tC and assuming one tC is roughly equivalent to 4 tCO_2 (Klein and Parry, 2007).

$$\text{CO}_2\text{T} = \frac{\text{HUR} \cdot \text{EPA42factor} \cdot \eta \cdot \text{SCCO}_2}{\text{ExR}} \tag{A9}$$

where CO_2T is the annualized benefit from reduction of CO_2 emission taxes [MM£/y], HUR is the Hot Utility Reduction [MMBtu/y], EPA42 factor is the EPA-42 emissions factor [MMTon CO_2/MMBtu], η is the equipment efficiency [–], $SCCO_2$ is the Estimated Social Cost of CO_2 [USD/tCO_2$] and ExR is an exchange rate average of equivalent $USD to 1 GBP (£).

Execution of the Retrofit Design Approach generates a retrofit portfolio, which includes all the proposed cost-effective changes together with their effects on profits, capital costs, CO_2 tax reductions and payback periods. These portfolios of results provide an important conceptual understanding with which process engineers can deal with process retrofit in confidence.

APPENDIX B. DATA AND INFORMATION FOR CASE STUDY

Table A1: Product specifications in the HCF plant

Product	Component	Unit	Specification
Ethane (C_2)	H_2S	ppm	≤50
	CO_2	% vol	≤0.03
	Methane	% vol	≤3.5%
	Ethane	% vol	≥93%
	Propane	% vol	≤4%
Propane (C_3)	Propane	% vol	≥98%
Propane–butanes (LPG)	Ethane	% vol	≤2.5
	Pentane	% vol	≤2
	Sulfur	ppm	≤140
Light naphtha (C_{5+})	Butanes	% vol	≤2
	Sulfur	ppm	≤140

Table A2: Available utilities

Utilities	Temperature (°C)	Cost (£/kW^{-1} y^{-1})
Fuel gas	280	120
High pressure steam	450	379
Medium pressure steam	360	358
Low pressure steam	180	242
Hot water	90	33
Cooling water	25–35	25
Propane refrigeration	−45	472
Electricity		300

Table A3: Raw material and products unit costs

Component	Type	Unit cost
Feed to de-ethanizer (cryogenic liquids I)	Raw material liquid phase	139.9 (£/Nm^3)
Feed to de-butanizer (cryogenic liquids II)	Raw material liquid phase	177.8 (£/Nm^3)
Ethane (C_2)	Product gas phase	0.109 (£/Nm^3)
LPG (C_3/C_4)	Product liquid phase	0.3 (£/kg)
Light naphtha (C_{5+})	Product liquid phase	287 (£/Nm^3)
Heavy naphtha (C_{6+})	Product liquid phase	338.4 (£/Nm^3)

Table A4: Parameters used in the simulated annealing algorithm

Annealing parameters	Value
Random number generator seed	1
Initial annealing temperature	100,000,000
Final annealing temperature	1.00000×10^{-05}
Markov chain length	30
Maximum number of iterations	25,000
Maximum consecutive failed chains	10
Maximum unsuccessful moves	300

Cooling parameter	0.01
Move acceptance criteria	Metropolis

Table A5: Probabilities of moves in the simulated annealing algorithm

Heat exchanger changes		Bypass changes	
Add heat exchanger	0.01	Add bypass	0.34
Delete heat exchanger	0.01	Add split	0.35
Delete heat spare exchanger	0.01	Delete bypass	0.2
Modify heat duty	0.5	Modify bypass	0.1
Reconfigure heat exchanger	0.47	Delete spare mixer	0.01

Change class		Heat exchanger reconfiguration	
Heat exchanger change	0.33	Re-sequence heat exchanger	0.5
Bypass change	0.33	Re-pipe heat exchanger	0.5
Utility temperature change	0.34		

REFERENCES

1. Ahmad, M.I., Zhang, N., Jobson, M., Chen, L., 2012. Multi-period design of heat exchanger networks. Chem. Eng. Res. Des. 90 (11), 1883–1895.
2. Asante, D.K., Zhu, X.X., 1996. An automated approach for heat exchanger network retrofit featuring minimal topology modifications. Comput. Chem. Eng. 20 (suppl.), S7–S12.
3. Carvalho, A., Gani, R., Matos, H., 2008. Design of sustainable chemical processes: systematic retrofit analysis generation and evaluation of alternatives. Process Saf. Environ. Prot. 86, 328–346.
4. Chen, C.L., Lin, C.Y., Lee, J.Y., 2013. Retrofit of steam power plants in a petroleum refinery. Appl. Therm. Eng. 61, 7–16.

5. Dhole, V.R., Linnhoff, B., 1993. Total site targets for fuel, co-generation, emissions and cooling. Comput. Chem. Eng. 17 (suppl.), 101–109.
6. Feng, X., Liang, C., 2013. Strategy for total energy system retrofit of a chemical plant. Chem. Eng. Trans. 35, 145–150.
7. Gadalla, M., Kamel, D., Ashour, F., Nour El din, H., 2013. A new optimization based retrofit approach for revamping an Egyptian crude oil distillation unit. Energy Proc. 36, 454–464.
8. Grossmann, I.E., 2013, March. Recent developments in the application of mathematical programming to process integration. In: Int. Process Integration Jubilee Conference, Gothenburg, Sweden.
9. Hernández-Enríquez, A., Tanco, M., Kim, J., 2011. Simulation-based process design and integration for the sustainable retrofit of chemical processes. Ind. Eng. Chem. Res. 50, 12067–12079.
10. Jain, S., Smith, R., Kim, J., 2012. Synthesis of heat-integrated distillation sequence systems. J. Taiwan Inst. Chem. Eng. 43, 525–534.
11. Jackson, J.R., Grossmann, I.E., 2002. High-level optimization model for the retrofit planning of process networks. Ind. Eng. Chem. Res. 41, 3762–3770.
12. Kemp, I., 2007. Pinch Analysis and Process Integration, 2nd ed. Elsevier, Oxford, UK.
13. Klein, R.J.T., Parry, M.L., 2007. Inter-relationships between adaptation and mitigation. In: Climate Change 2007: Impacts, Adaptation and Vulnerability – Contribution of Working Group II to the Fourth Assessment Report of the Intergovernmental Panel on Climate Change. Cambridge University Press, Cambridge, U.K..
14. Klemes, J., Friedler, F., Bulatov, I., Varbanov, P., 2010. Sustainability in the Process Industry: Integration and Optimization. McGraw Hill, New York, US.
15. Krajnc, D., Glavic, P., 2009. Assessment of different strategies for the co-production of bioethanol and beet sugar. Chem.

Eng. Res. Des. 87, 1217–1231.

16. Koch-Glitsch, 2013. www.koch-glitch.com (accessed 20.09.13). Kokossis, A.C., Floudas, C.A., 1990. Optimization of complex reactor networks – I. Isothermal operation. Chem. Eng. Sci. 45 (3), 595–614.

17. Kokossis, A.C., Floudas, C.A., 1994. Optimization of complex reactor networks – II. Non-isothermal operation. Chem. Eng. Sci. 49 (7), 1037–1051.

18. Linnhoff, B., Hindmarsh, E., 1983. The pinch design and method for heat exchanger networks. Chem. Eng. Sci. 38 (5), 745–763.

19. Myers, R.H., Montgomery, D.C., Vinning, G.G., Borror, C.M., 2004. Response surface methodology: a retrospective and literature survey. J. Qual. Technol. 36, 53–78.

20. Polley, G.T., Tamakloe, E., Picon Nunez, M., Ishiyama, E.M., Wilson, D.I., 2013. Applying thermo-hydraulic simulation and heat exchanger analysis to the retrofit of heat recovery systems. Appl. Therm. Eng. 51, 137–143.

21. Reddy, C.C.S., Rangaiah, G.P., Long, L.W., Naidu, S.V., 2013. Holistic approach for retrofit design of cooling water networks. Ind. Eng. Chem. Res. 52, 13059–13078.

22. Shah, P.B., Kokossis, A.C., 2002. New synthesis framework for the optimization of complex distillation systems. AIChE J. 48, 527–550.

23. Simon, L.L., Osterwalder, N., Fisher, U., Hungerbuhler, K., 2008. Systematic retrofit method for chemical batch processes using indicators, heuristics and process models. Ind. Eng. Chem. Res. 47, 66–80.

24. Smith, R., 2005. Chemical Process Design and Integration. Wiley, Chichester, UK.

25. Smith, R., Jobson, M., Chen, L., 2010. Recent development in the retrofit of heat exchanger networks. Appl. Therm. Eng. 30 (16), 2281–2289.

26. Smith, R., 2013, March. Recent development in process

integration design techniques. In: Int. Process Integration Jubilee Conference, Gothenburg, Sweden.
27. Smith, R., Linnhoff, B., 1988. Distillation columns with three products. Chem. Eng. Res. Des. 66, 195.
28. Tahouni, N., Bagheri, N., Twofighi, J., Panjeshahi, M.H., 2013. Improving energy efficiency of an olefin plant – a new approach. Energy Convers. Manage. 76, 453–462.
29. Timmerhaus, K., Peters, M.S., 2003. Plant Design and Economics for Chemical Engineers, 5th ed. McGraw-Hill, New York, USA.
30. US EPA (Env. Protection Agency), 2006. Emission Factor Documentation for AP-42 Section 1.4 – Natural Gas Combustion (EPA-R09-OAR-2006-0635-0013). US Environmental Protection Agency: Research Triangle.
31. Wang, Y., Smith, R., 2013. Retrofit of a heat-exchanger network by considering heat-transfer enhancement and fouling. Ind. Eng. Chem. Res. 52, 8527–8537.

Chapter 6

Production of Hydrocarbon Liquid by Thermal Pyrolysis of Paper Cup Waste

Bijayani Biswal, Sachin Kumar, and R. K. Singh

Department of Chemical Engineering, National Institute of Technology, Rourkela, Orissa 769008, India

ABSTRACT

The paper cup waste was pyrolysed in a stainless steel semibatch reactor at a temperature range of 325°C to 425°C and at a heating rate of 20°C min^{-1} with an aim to study the physical and chemical characteristics of the obtained hydrocarbon liquid and to determine its feasibility as a commercial fuel. The maximum liquid yield was 52% at 400°C. The functional groups present in the liquid are aldehydes, ketones, carboxylic acids, esters, alkenes, and alkanes.

It was found that the pyrolytic liquid contains around 18 types of compounds having carbon chain length in the range of C_6–C_{20}. The obtained liquid can be used as valuable chemicals feedstock.

INTRODUCTION

Urbanization is an important determinant of both the quantity and the type of fuel used in developing countries. In general, urbanization leads to higher levels of energy consumption, also accompanied with increases in income levels. Therefore, there is a shift from traditional to commercial fuels. Several other factors that contribute to this trend include decline in access to biomass fuels, inconvenience of transportation and storage of biomass fuels, and improvement unavailability of commercial fuels in urban areas [1]. The disposal of solid biomass and waste is becoming an enormous problem because they are very difficult and costly to manage. Pyrolysis has proved itself to be a new type of solid biomass and waste utilization technique that transforms biomass and waste material of low-energy density into bio-oil of high-energy density and recover higher value chemicals. Paper cups used as coffee or cold drinks cups are accumulating as wastes on the earth surface at a rapid rate. Considering only America, 14.4 million disposable paper cups are used for drinking coffee each year. Placed end-to-end, these cups would wrap around Earth 55 times and weigh around 900 million pounds.

Most paper cups are designed for a single use and then disposed or recycled. One paper cup represents 4.1 g equivalent petrol with a production cost 2.5 times higher than plastic cups. Paper cups are not specifically recycled. They come under regular waste and burnt or put on landfills. Recycling paper cups is difficult because of its composition as a complex of paper and paraffin. Hence, they need about 150 years (same as plastics) to degrade because of their plastic foil [2]. The paper cups for hot drinks are produced from wood pulp (cellulose) and polyethylene plastic film, made out of petrol or paraffins, to improve its water resistivity and resistance to heat. They have a coating of 8–18 g/m² on one side. Cups for cold drinks have 6–15 g/m² on the top side and 8–18 g/m² on the reverse side [3]. A basic hot beverage cup is typically made of 95% (by wt) of paper and 5% (by wt) of polyethylene for coating. In cold drink cup, polythene used is 10% and fiber is 90%. The paper used is produced from valuable "bleached kraft" fibres [3]. Mixed paper wastes (MPW) represent a valuable source of energy. Hence, it is studied to determine the quantity of energy obtained from waste of known amount and composition. For a waste to become an energy system, the heating value of waste is one of the important characteristics that determine the energy obtainable from wastes [4]. Mixed paper wastes comprising a mixture of newspaper, cardboard, kraft, beverage and milk boxes, boxboard, tissue, colored office paper, white office paper, envelopes, treated paper (NCR), and glossy paper were studied for their combined calorific value as shown in Table 1 [4].

Table 1: Calorific value of different paper waste [4]

Type of paper	Mean gross calorific value (Btu/lb)
Newspaper	7540
Cardboard	6907
Kraft	6897
Beverage and milk boxes	6855
Boxboard	6703
Tissue	6518
Colored office paper	6348
White office paper	6234
Envelopes	6160
Treated paper (NCR)	5983
Glossy paper	6370
Mixed	6477

The expected calorific value of any unknown sample of mixed paper can be calculated by using the calorific values of each individual category of mixed waste paper (MWP) and the weight fraction of each in a MWP sample [4]. As paper cups are low-density polyethylene (LDPE) coated, so the study of pyrolysis of waste paper and LDPE could be of some help. Li et al. studied the influence of pyrolysis temperature and heating rate on yield of pyrolysis products from waste paper and concluded that the maximum biooil yield of 49.13% was achieved at a temperature around 420°C with heating rate of 30°C min^{-1}. The results of spectroscopic and chromatographic analysis show that bio-oil contained many different chemical classes, and there are four main different compounds in bio-oil: anhydrosugars, carboxyl compounds, carbonyl compounds, and aromatic compounds [5]. Pyrolysis of uncoated printing and writing paper was carried out by Wu et al. in a TGA reaction system at a constant heating rate of 5 K min^{-1} and in a nitrogen environment. The gaseous products investigated included nonhydrocarbons (H_2, CO, CO_2, and H_2O) and hydrocarbons (C_1–C_3, C_4, C_5, C_6, benzene, C_{10}–C_{12}, levoglucosan, C_{13}–C_{15}, and C_{16}–C_{18}). The cumulated masses and the instantaneous concentrations of gaseous products were obtained under the experimental conditions. The yields of non-hydrocarbon gases and of hydrocarbons were about 10.46 and 0.49% at 623 K, 33.68 and 0.89% at 700 K, 64.52 and 1.05% at 788 K, and 79.10 and 1.63% at 938 K, respectively. Since the synthetic gases (CO, CO_2, H_2O, HCs) contained a high calorific value, their use as marketable fuels gently supported the importance for resource recycling of the uncoated printing and writing paper [6]. Shah et al. have pyrolysed waste LDPE in a home-assembled batch reactor under atmospheric pressure using a wide range of acidic and basic catalyst like silica, calcium carbide alumina, magnesium oxide, and homogenous mixture of silica and alumina. CaC_2 proved advantageous on basis of reaction time while the efficiency of conversion to liquid was more for SiO_2 at optimum conditions. Hence, these two can be suitably used for catalytic pyrolysis of polyethylene (optimum weight—1 g/5 g of LDPE). Oxide containing catalyst could be best suited for selective conversion into polar and aromatic compounds while CaC_2 could be adopted for selective conversion into aliphatic products [7]. The liquid product obtained from catalytic pyrolysis was characterized by physical and chemical tests. Physical tests include density, specific gravity, API gravity, viscosity, aniline point, flash point, and gross calorific value. These were determined according to IP and ASTM standard methods for fuel values. The liquid fraction obtained is comparable with the standard results of physical tests for gasoline, kerosene and diesel fuel oil. Chemical tests like bromine water and $KMnO_4$ tests the presence of mixture of olefin, and aromatic compounds can be obtained. Components in the oil mixture were separated by column chromatography and fractional distillation followed by characterization with FTIR spectroscopy [7].

The objective of the present work is to optimize the liquid fuel production from waste paper cups by thermal pyrolysis in a semibatch reactor. This work also reports on the characterization of the liquid fuel using FTIR, GC-MS for composition, and other standard methods for studying different physical properties.

MATERIALS AND METHODS

Paper cups waste has been collected from NIT campus area in Rourkela, Orissa (India). The cups were cut into small square-shaped pieces (about 1 cm side). The proximate analysis of paper cups waste and char obtained after pyrolysis was done by ASTM D3173-75 and ultimate analysis was done using CHNS analyzer (elementar vario EL cube chnso). Calorific value of the raw material and the obtained char was done by ASTM D5868-10a.

Thermogravimetric analysis of the waste paper cup sample was carried out with a shimadzu DTG-60/60H instrument. A known weight of the sample was heated in a silica crucible at a constant heating rate of 25°C/min operating in a stream of nitrogen with a flow rate of 25 mL/min from 32°C to 700°C.

The pyrolysis setup consists of a semi batch reactor made of stainless steel tube (length, 145 mm, internal diameter, 37 mm and outer diameter, 41 mm) sealed at one end and an outlet tube at the other end. The reactor is heated externally by an electric furnace, with the temperature being measured by a Cr-Al: K type thermocouple fixed inside the reactor, and temperature is controlled by external PID controller as shown in the previous study [8]. 15 g of waste paper cup sample was loaded in each pyrolysis reaction. The condensable liquid products/wax were collected through the condenser and weighed. After pyrolysis, the solid residue left out inside the reactor was weighed. Then, the weight of gaseous/volatile product was calculated from the material balance. Reactions were carried out at different temperatures ranging from 325–425°C with a temperature difference of 25°C.

Fourier transform infrared spectroscopy (FTIR) of the pyrolytic oil obtained at optimum condition was taken with a Perkin-Elmer Fourier-transformed infrared spectrophotometer with resolution of 4 cm^{-1}, in the range of 400–4000 cm^{-1} to know the functional group composition. The components of liquid product were analyzed using GC-MS-QP 2010 (shimadzu). The GC conditions, column oven temperature progress, column used, and MS conditions are given in Table 2. Physical properties such as density, specific gravity, viscosity, Conradson carbon, flash point, fire point, pour point, cloud point, calorific value, sulphur content, and cetane index of the liquid were determined using the standard test methods. The water content in liquid product was determined by the Karl Fischer method and was separated by the gravity separation of the liquid.

Table 2: GC-MS conditions

Instrument	GC-MS-OP 2010 [SHIMADZU]	
GC conditions		
Column oven temperature	70°C	
Injection mode	Split	
Injection temperature	200°C	
Split ratio	10	
Flow control mode	Linear velocity	
Column flow	1.51 mL/min	
Carrier gas	Helium 99.9995% purity	

Column oven temperature progress		
Rate	Temperature (°C)	Hold time (min)
—	70	2
10	300	7.0 (32 min total)
Column: DB-5		
Length	30.0 m	
Diameter	0.25 mm	
Film thickness	0.25 m	
MS conditions		
Ion source temperature	200°C	
Interface temperature	240°C	
Start m/z	40	
End m/z	1000	

RESULT AND DISCUSSION

Proximate and Ultimate Analysis of Paper Cup Waste and Obtained Char

The proximate and ultimate analyses of waste paper cup sample are shown in Table 3. The volatile matter is 52% in the proximate analysis which is drastically reduced to 12% after pyrolysis. It indicates high conversion of biomass to liquid fuels. As a result of decrease in volatile matter content, fixed carbon of material increased significantly which means there is less liberation of fixed carbon. Ultimate analysis presented in Table 3showed significant variation in carbon and oxygen content, whereas there were slight variations in hydrogen, nitrogen, and sulphur content.

Table 3: Proximate and ultimate analysis of paper cup waste and obtained char after pyrolysis

	Paper cup	Paper cup char
Proximate analysis (wt%)		
Moisture content	0	0
Volatile matter	52	12
Ash content	2	8
Fixed carbon	46	80
Ultimate analysis (wt%)		
C	46.7	77.7
H	6.7	4.9
N	2.12	5.33
S	0	0
O	44.4	12.07
C/H molar Ratio	0.57	1.32
C/O molar Ratio	1.4	8.63
Empirical formula	$C_{1.4}H_{2.41}N_{0.05}O$	$C_{0.017}H_{0.012}N_{0.01}O_{0.0019}$
Gross calorific value (MJ/Kg)	20.1	25.4

TGA and DTG Analysis of Paper Cup Waste

Thermogravimetric analysis (TGA) is a very useful thermal analysis technique to investigate the thermal stability of a material, or to investigate its behavior in different atmospheres (e.g., inert or oxidizing). TGA applied for the study of thermal stability/degradation of paper cup waste in various ranges of temperature. The characteristic parameters of de-volatilization are presented in this section. The TGA plot of waste paper cups at a heating rate of 25°C/min under nitrogen atmosphere is shown in Figure 1. The TGA of paper cup waste shows that the active pyrolytic zone was in between the temperature range of 260–410°C. In this case, the first stage decomposition represents the evaporation of moisture contents; second decomposition indicates the formation of volatiles mainly CO and CO_2. During the third stage, the pyrolysis residue slowly decomposed, with the weight-loss velocity becoming smaller and smaller, and the residue ratio tends to be constant at the end the decomposition of hydrocarbon. A three-stage weight loss is observed. In the initial or first stage decomposition, 7.39% weight loss was observed which represents the removal of moisture content. In active pyrolytic zone or second stage decomposition, 70.42% weight loss was observed, and in the third stage decomposition, 22.19% weight loss was observed. Due to high decomposition rate per unit time, the rapid decomposition zone or scecond stage of decomposition is treated as active pyrolytic zone. During the second stage, the intermolecular associations and weaker chemical bonds are destroyed [9–13]. The side aliphatic chains may be broken, and some small gaseous molecules are produced at the lower temperature. During the third stage at higher temperature chemical bonds are broken and the parent molecular skeletons are destroyed. As a result, the larger molecule decomposes to form smaller molecules.

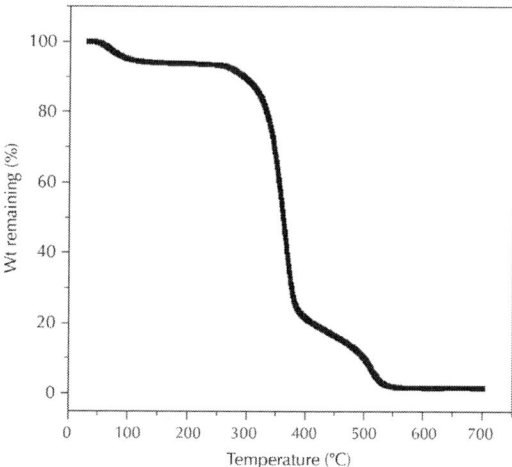

Figure 1: TGA plot of paper cup waste.

Differential thermogravimetry (DTG) (Figure 2) curve for paper cup waste contains one major peak; this indicates that there is one key degradation step in Figure 2. The dominant peak was at a temperature from 270°C to 405°C where the maximum decomposition occurred. Similarly, the active pyrolysis zone for glossy paper waste was 274–361°C where the maximum weight loss occurred [14]. Similar trend has been observed for the newspaper where the second decomposition of the newspaper occurred between 291°C and 429°C, representing a 63.2% weight loss [15].

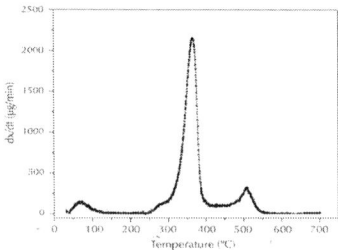

Figure 2: DTG plot of paper cup waste.

Effect of Temperature on Product Distribution and Reaction Time

The pyrolysis of paper cup waste yielded three different products, that is, oil, gas, and residue. The distributions of these fractions are different at different temperatures and are shown in the Table 4.

Table 4: Distribution of different fractions at different temperatures in thermal pyrolysis of paper cup waste

Temperature (°C)	Oil (wt.%)	Gas/volatile (wt.%)	Char (wt.%)	Reaction time (min.)
325	30.33	26.67	43	24
350	39.23	21.27	39.5	20
375	47.13	16.87	36	18
400	52.53	13.87	33.6	15
425	42.33	26.87	30.8	8

The oil and gas/volatiles constituted major product as compared to the solid residue fractions. The condensable product obtained at low temperature (325°C and 350°C) was low viscous liquids. With increase in temperature, the liquid was viscous, and the maximum liquid product was 52% at the temperature 400°C. The recovery of condensable fraction increased with gradual increase of temperature. Similarly, for the waste paper, a maximum of 49.13% of the liquid product has been obtained at pyrolysis temperature of 420°C [5]. The maximum yield of char 43% was obtained at the temperature 325°C. With the increasing temperature, the char yield is decreased. This decrease could be due to secondary decomposition of char residue or not decomposed materials [16]. The yield of gas was decreased with increasing pyrolysis temperature up to 400°C and then increases. The highest yield of gas 26.87% was obtained at a pyrolysis temperature of 425°C. This is due to the secondary decompositions of the char, and secondary cracking of pyrolysis vapors may enrich the contents of the gas product at higher temperature [16–18]. The effect of temperature is shown in Figure 3. The pyrolysis reaction rate increased and reaction time decreased with increase in temperature.

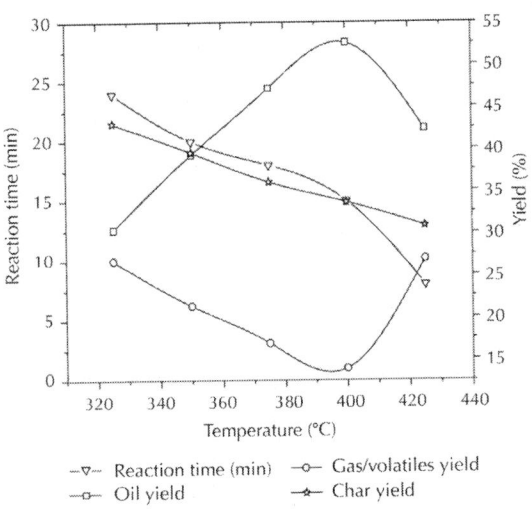

Figure 3: Effect of temperature on product yield and reaction time.

CHARACTERIZATION OF THE LIQUID PRODUCT

FTIR of the Oil Sample Obtained at 400°C

Fourier transform infrared spectroscopy (FTIR) is an important analysis technique which detects various characteristic functional groups present in oil. Figure 4 shows the FTIR spectra of paper cup waste pyrolytic oil. The O–H stretching vibrations at frequency 3409 cm^{-1} indicate the presence of alcohol. The presence of alkanes is detected at 2851 cm^{-1} with C–H stretching vibrations. C=O stretching vibrations at 1714 cm^{-1} show the presence of aldehydes, ketones, carboxylic acids, esters. The presence of alkenes was detected by C=C stretching vibrations at 1644 cm^{-1}. The presence of alcohols, ethers, carboxylic acids, and esters is detected by C–O stretching vibrations at 1057 cm^{-1}. C–H bending vibrations at 925 cm^{-1} indicate the presence of alkenes, and the C–H bending vibrations at frequency 811 cm^{-1} indicate the presence of phenyl ring substitution bands. The results were found consistent when compared with the results of GC-MS.

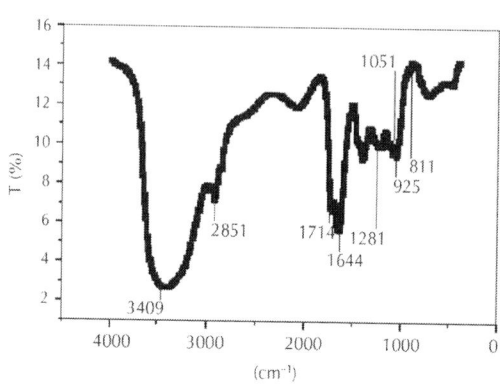

Figure 4: FTIR spectrometry of paper cup waste pyrolytic oil.

GC-MS of the Oil Sample

The GC-MS analysis of the pyrolytic oil sample (Figure 5) is summarized in Table 5. It has been observed that the pyrolytic oil contains around 18 compounds. Taking into account the area percentage, the highest peak areas of total ion chromatogram (TIC) of the compounds were 2-furancarboxaldehyde, 2-furaldehyde, oleanitrile, hexadecanoic acid, hexadecanenitrile, methyl cyclopentenolone, and eicosane. The components present in paper cup waste pyrolytic oil are mostly furan derivatives of ketones, phenols, and the aliphatic hydrocarbons (alkanes) with carbon number C_6-C_{20}.

Table 5: GC-MS of paper cup waste pyrolytic oil

Compound	IUPAC name	Area%	Formula
1-Acetoxy-2-propionoxyethane	Dimethyl 2,2-dimethylpropanedioate	3.62	$C_7H_{12}O_4$
2-Furancarboxaldehyde, 5-methyl-	Benzene-1,3-diol	10.36	$C_6H_6O_2$
Phenol	Hydroxybenzene	2.34	C_6H_5OH
1,2-Cyclopentanedione, 3-methyl-	Cyclohexa-3,5-diene-1,2-diol	5.06	$C_6H_8O_2$
2,3-Dihydro-5-hydroxy-6-methyl-4H-pyran-4-one	2,3-dihydroxy-4-methylcyclopent-2-en-	3.27	$C_6H_8O_3$
Oxetane, 2-Propyl	4-methylpentan-2-one	2.85	$C_6H_{12}O$
Maltol	Prop-2-enoyl prop-2-enoate	5.82	$C_6H_6O_3$
2,3-Dihydro-3,5-dihydroxy-6-methyl-4H-pyran-4-one	Cyclobutane-1,1-dicarboxylic acid	1.17	$C_6H_8O_4$
1,4:3,6-Dianhydro-.alpha.-d-glucopyranose	Cyclobutane-1,1-dicarboxylic acid	12.26	$C_6H_8O_4$

2-Furancarboaldehyde, 5-(hydroxymethyl)-	Prop-2-enoyl prop-2-enoate	18.20	$C_6H_6O_3$
Tetradecane	Tetradecane	0.72	$C_{14}H_{30}$
Pentadecane	7-ethyltridecane	1.51	$C_{15}H_{32}$
3-Heptadecene, (Z)-	4,8,12-trimethyltetradec-1-ene	1.55	$C_{17}H_{34}$
Eicosane	Icosane	8.54	$C_{20}H_{42}$
Hexadecanenitrile	(E)-N-[(E)-oct-1-enyl]oct-1-en-1-amine	1.08	$C_{16}H_{31}N$
Hexadecanoic acid	2-(11-methyldodecoxymethyl)oxirane	3.59	$C_{16}H_{32}O_2$
Oleanitrile	2-decyl-1,3,3a,4,7,7a-hexahydroisoindole	3.86	$C_{18}H_{33}N$
Octadecanoic acid	1-Acetoxyhexadecane	1.84	$C_{18}H_{36}O_2$

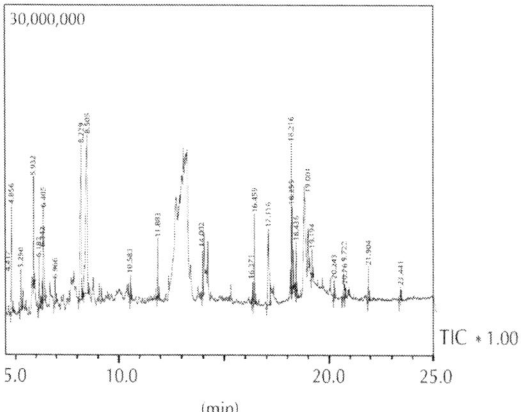

Figure 5: GC-MS plot of paper cup waste pyrolytic oil.

Physical Properties of Oil Sample

Table 6 shows the results of physical property analysis of oil obtained from pyrolysis of paper cup waste. The appearance of the oil is dark brownish free from visible sediments.

Table 6: Physical properties analysis of paper cup waste pyrolytic oil

Properties	Pyrolytic oil	Standard test methods
Density at 15°C in kg/m^3	1.0136	ASTM D1298-99
Specific gravity at 15°C/15°C	1.0145	ASTM D1298-99
Kinematic viscosity at 40°C in centistokes	0.80	ASTM D445-11
Flash point	64°C	ASTM D6450-05(2010)
Pour point	Below −12°C	ASTM D5853-09
Cloud point	−4°C	ASTM D1310-01(2007)
Sulphur content	Nil	IS:1448 P:33
Conradson carbon residue	0.76%	ASTM D189-06(2010)e1
Water content	9.1%	Karl Fischer Method
Gross calorific value in MJ/Kg	23	ASTM D5468-02(2007)
Distillation		ASTM D2887-08
Initial boiling point	187°C	
Final boiling point	369°C	

From comparison with other transportation fuels as shown in Table 7, the density and viscosity of liquid product can be modified by blending it with commercial transportation fuels. The flash point of the liquid product is in a comparable range, and pour point is minus 12°C which will not cause any trouble in most of the regions, but in colder regions with subzero climates it may lead to freezing problems. Paper cup waste pyrolytic oil has GCV of 23 MJ/Kg which is less as compared to that of gasoline and diesel; therefore, this liquid product is a very poor engine fuel. From the distillation report of the oil it is observed that the boiling range of the oil is 187–369°C, which infers the presence of a mixture of different oil components such as gasoline, kerosene, and diesel in the oil. From this result, it is observed that this hydrocarbon liquid could be possible feedstock for further upgrading or use of lighter compounds as a diesel fuel.

Table 7: Comparison of fuel properties of paper cup waste pyrolytic oil with transportation fuels

Fuel properties	Specific gravity 15°C/15°C	Kinematic viscosity at 40°C (cst)	Flash point(C)	Pour point(C)	GCV(MJ/Kg)	IBP(C)	FBP(C)	Chemical formula
Paper cup Pyrolytic oil	1.0145	0.80	64	< −12°C	23	187	369	C_6–C_{18}
Gasoline [19]	0.72–0.78	—	−43	−40	42–46	27	225	C_4–C_{12}
Diesel [20]	0.82–0.85	2–5.5	53–80	−40 to −1	42–45	172	350	C_8–C_{25}
Bio-diesel [20]	0.88	4–6	100–170	−3 to 19	37–40	315	350	C_{12}–C_{22}

CONCLUSIONS

Thermal pyrolysis of paper cup waste was performed in a semi batch reactor at a temperature range from 325°C to 425°C and at a heating rate of 20°C/min. The maximum liquid yield was 52% at temperature 400°C; volatile products are mainly obtained at low temperature. Reaction time decreases with increase in temperature. The functional groups present in the pyrolytic oil are aldehydes, ketones, carboxylic acids, esters, alkenes, and alkanes. It was found that the pyrolytic oil contains around 18 compounds having carbon chain length in the range of C_6–C_{18}. The physical properties of pyrolytic oil obtained were in the range of other pyrolytic oils and poor quality fuels. A simple batch pyrolysis method can convert paper cup waste to liquid hydrocarbons with a significant yield which varies with temperature.

REFERENCES

1. D. Oleg and C. Ralph, "Trends in consumption and production: household energy consumption," DESA Discussion paper no. 6, 1999.
2. "Comparative Study: identifying mug, plastic cup, biodegradable and compostable cups, and paper cups environmental qualities," http://www.eco-collectoor.fr/.
3. http://www.environmentalgraffiti.com/waste-and-recycling/news-newest-alternative-fuel-source-paper-cups.
4. Ucuncu, "Energy recovery from mixed paper wastes," Final Report to Sunshares, Durham, NC, USA.
5. L. Li, H. Zhang, and X. Zhuang, "Pyrolysis of waste paper: characterization and composition of pyrolysis oil," Energy Sources, vol. 27, no. 9, pp. 867–873, 2005.
6. C.-H. Wu, C.-Y. Chang, and C.-H. Tseng, "Pyrolysis products of uncoated printing and writing paper of MSW," Fuel, vol. 81, no. 6, pp. 719–725, 2002.

7. J. Shah, M. R. Jan, F. Mabood, and F. Jabeen, "Catalytic pyrolysis of LDPE leads to valuable resource recovery and reduction of waste problems," Energy Conversion and Management, vol. 51, no. 12, pp. 2791–2801, 2010.
8. S. Kumar and R. K. Singh, "Recovery of hydrocarbon liquid from waste high density polyethylene by thermal pyrolysis, Braz," Chemical Engineering Journal, vol. 28, no. 04, pp. 659–667, 2011.
9. D. Jinno, A. K. Gupta, and K. Yoshikawa, "Determination of chemical kinetic parameters of surrogate solid wastes," Journal of Engineering for Gas Turbines and Power, vol. 126, no. 4, pp. 685–692, 2004.
10. P. W. Chan, A. Atreya, and R. B. Howard, "Determination of pyrolysis temperature for charring materials," Proceedings of the Combustion Institute, vol. 32, pp. 2471–2479, 2009.
11. P. Raman, W. P. Walawender, L. T. Fan, and J. A. Howell, "Thermogravimetric analysis of biomass, devolatilization studies on feedlot manure," Industrial & Engineering Chemistry Process Design and Development, vol. 20, pp. 630–636, 1981.
12. C. Liu, J. Yu, X. Sun, J. Zhang, and J. He, "Thermal degradation studies of cyclic olefin copolymers," Polymer Degradation and Stability, vol. 81, no. 2, pp. 197–205, 2003.
13. Demirbas, "The influence of temperature on the yields of compounds existing in bio-oils obtained from biomass samples via pyrolysis," Fuel Processing Technology, vol. 88, no. 6, pp. 591–597, 2007.
14. J. K. Modh, S. A. Namjoshi, and S. A. Channiwala, "Kinetics and pyrolysis of glossy paper waste," International Journal of Engineering Research and Applications, vol. 2, pp. 1067–1074, 2012.
15. M. N. A. Bhuiyan, M. Ota, K. Murakami, and H. Yoshida, "Pyrolysis kinetics of newspaper and its gasification," Energy Sources A, vol. 32, no. 2, pp. 108–118, 2010.
16. P. A. Horne and P. T. Williams, "Influence of temperature on

the products from the flash pyrolysis of biomass," Fuel, vol. 75, no. 9, pp. 1051–1059, 1996.

17. L. Fagbemi, L. Khezami, and R. Capart, "Pyrolysis products from different biomasses: application to the thermal cracking of tar," Applied Energy, vol. 69, no. 4, pp. 293–306, 2001.

18. S. Galvagno, S. Casu, T. Casabianca, A. Calabrese, and G. Cornacchia, "Pyrolysis process for the treatment of scrap tyres: preliminary experimental results," Waste Management, vol. 22, no. 8, pp. 917–923, 2002.

19. "Petroleum Product Surveys, Motor Gasoline, Summer, Winter 1986/1987," National Institute for Petroleum and Energy Research, 1986.

20. J. Tuttle and T. V. Kuegelgen, Biodiesel Handling and Use Guidelines, National Renewable Energy Laboratory, 3rd edition, 2004.

Chapter 7

Effect of Soil Texture on Remediation of Hydrocarbons-contaminated Soil at El-Minia District, Upper Egypt

Th. Abdel-Moghny[1], Ramadan S. A. Mohamed[2],
E. El-Sayed[2], Shoukry Mohammed Aly[3], and
Moustafa Gamal Snousy[2]

[1]Applications Department, Egyptian Petroleum Research Institute, Ahmed El-Zomer, Nasr City, Cairo, Egypt

[2]Geology Department, Faculty of Science, El-Minia University, El-Minia, Egypt

[3]Petrotreat Co., Egypt

ABSTRACT

Soils polluted by waste lubricant oils may affect the hydrosphere compromising the quality of drinking water resources and threatening the aquatic ecosystems. The objective of this study focused to remove waste-lubricant oils from different polluted sites in El-Minia governorate. In this respect some samples were collected from four different industrial sites and identified as sand, loamy sand, clay loam and loam. Then the field conditions were simulates using two experimental models packed with contaminated soil. The remediation processes carried out in both models using surfactant enhanced by air injection then by water washing. The parameters such as soil type, soil heterogeneity, time and washing process was investigated. The results indicated that the high efficiency of oil removal is obtained from sand where the clay loam gives the worst results. The results also reveal that, the high flushing and washing duration time can be attributed to the high percentage of mud in some sites over other sites. This means that the performance of surfactant flushing/water washing can be adversely affected by geologic heterogeneity. Finally, it's suitable to use pressurized liquid technologies in heterogeneous media, but cleanup times will be longer and more difficult than for the other similar homogeneous media.

INTRODUCTION

Throughout the world, subsurface contamination has become a widespread and pervasive problem. A major problem in the soil or groundwater remediation is the removal of hydrophobic organic compounds. Nonaqueous phase liquids (NAPLs) usually enter the unsaturated zone as discrete liquid phases, which move due to gravitational and capillary forces [1]. They frequently enter groundwater systems after they have been spilled on the surface and pass through the unsaturated zone (Figure 1). The major organic chemical waste categories include organic aqueous waste (e.g., pesticides), organic liquids (e.g., chlorinated solvents), oils (e.g., different fuels and fuel additives), and sludges or solids containing organic compounds. The most common local source of soil and water contamination by petroleum hydrocarbons are industrial plants, land disposal sites of danger residues, petrol stations, car service stations, and vehicle accidents. The total petroleum hydrocarbons include saturated alkanes, aromatic hydrocarbons, fuel oxygenated additives (e.g., methyl t-butyl ether (MTBE), ethanol, butanol), and other compounds containing sulfur or nitrogen. These compounds are harmful or even toxic to the growth and development of plants and animals, being a source of long-term water and air pollution. They are also dangerous to the human health [2]. Accidental surface release and improper disposal of petroleum products (e.g., jet fuel, refinery wastes, diesel, lubricating used oil, etc.) and volatile organic solvents are recognised as two of the most widespread causes of groundwater contamination by chemical compounds. Flooding and/or accidents, oil from the waste pit may spread to the surrounding fields causing pollution [3].

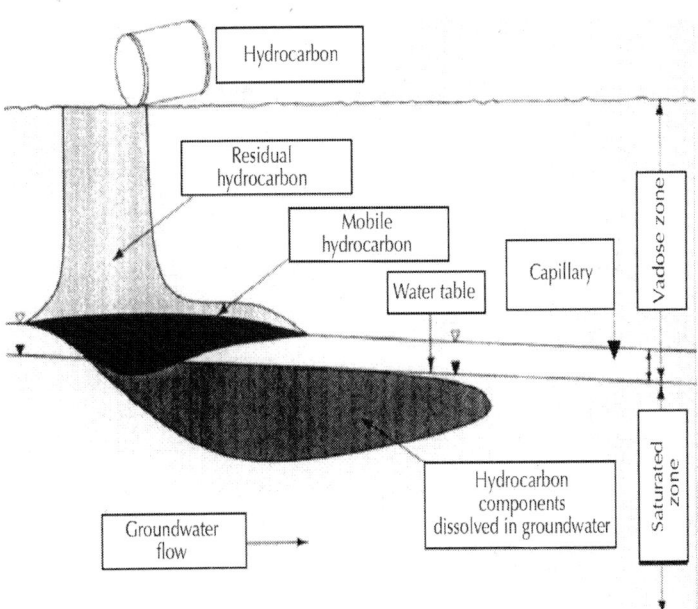

Figure 1: Illustration of the subsurface distribution of an NAPL spill [4].

Nonaqueous phase liquids (NAPLs) that are less dense than water, including many petroleum products or fuels, such as gasoline, heating oil, lubricating oil, kerosene, jet fuel, and aviation gas, are termed light nonaqueous phase liquids (LNAPLs) and these commonly collect and pool at, or above, the water table. The other type of NAPL is denser than water and named dens nonaqueous phase liquids (DNAPLs) include chlorinated hydrocarbons or chlorohydrocarbon such as carbon tetrachloride, 1,1,1-trichloroethane, chlorophenols, chlorobenzenes, tetrachloroethylene, and polychlorobiphenyls (PCBs). The latter are thus particularly difficult to remediate [5]. Matters become even more complicated since these contaminants are often mixed with metals. A considerable amount of hydrocarbon oils can be held in voids in the soil in the form of residual saturation and can lead to long-term contamination of groundwater through the action of rain water, if not removed in time [6]. During the transportation of NAPLs through the subsurface, a portion of the organic phase also retained within the pores of the soil matrix as an immobile ganglia or globules due to interfacial forces.

Physical Properties of Soil

The procedure selected to contain spills on land will vary with the amount and type of oil spilled, the type of soil, and the terrain. Less viscous oil and more porous soil will allow greater and more rapid penetration and lateral migration in the soil. Groundwater is very susceptible to contamination, unless protected by a low permeability layer such as clay. The organic contaminants like petroleum hydrocarbons, halogenated organic compounds, or other organic compounds are bind strongly inside the soil matrix and present for long time at the contaminated sites. Knowledge of physical properties of soil is most important for designing the parameters of remediation process. The mechanisms of interaction between the soil and contaminants are also important to know. Soil can be defined as loose material composed of weathered rock, other minerals, and also partly decayed organic matter, that covers large parts of the land surface [7]. The soil is composed of three phases: solid, liquid, and gas phases. The soil components include about 50% by volume mineral particles, 25% water, 20% air, and 5% organic matter. With the exception of a few organic soils, the bulk of soil material is mineral in character and has been derived from solid geological deposits [8]. The mineral constituents of the soil are represented by the particles of widely varying size, shape, and chemical composition. Three groupings of soil particles are in common use, namely, sand, silt, and clay. The groups are subdivided according to requirements.

Long-Term Contamination

The NAPL that remains in the unsaturated zone is an important source of contamination because it is dissolved by (1) the passing recharge water and (2) the passing groundwater as the water table rises. Such sources of contamination can last for many years and contaminate large volumes of groundwater. However, in addition to these pathways, contaminants also can be transported through the unsaturated zone. This transport pathway may spread the contaminants over a much broader area of the aquifer. In recent years, there has been an increasing interest in the remediation of NAPLs source zones [9]. Because of the low solubility of hydrophobic organic compounds in water, the residual organic phase usually represents a long-term contamination source for soil and groundwater. Owing to the tendency of contaminants to tightly bind or absorb onto the soil particles. Subsurface contamination by the organic compounds is a complex process and difficult to treat due to many reasons like the tendency of adsorption of contaminants onto the soil matrix, low water solubility, and limited rate of mass transfer for biodegradation, and so on [10]. As many organic compounds have low solubility in water, so they may leach from the soil for a longer period of time and thus ultimately become a continuous source of the soil and groundwater contamination. The use of surfactants can improve the mobility of hydrocarbon contaminants in soil-water systems by solubilising adsorbed hydrocarbons through incorporation in surfactant micelles [11]. This work aims to remove NAPL (waste-lubricant oils) polluted soils and collected from different industrial sites in El-Minia governorate. The air flushing as distributive system enhanced by surfactant and different water flooding cycles were used in this study. Two models were designed to simulate the injection wells and treatment tanks. Nonionic surfactant solutions Nonyl phenol ethoxylate (NPEO$_9$) were used at constant concentrations (3%) beyond its critical micelle concentration. The results discussed based on soil heterogeneity to select the suitable remedial techniques (in situ or ex situ).

Samples Locations

The design of optimal remediation schemes often requires some "prediction" of the distribution of contaminants within the subsurface over time. These predictions can be used to evaluate different remediation scenarios. Two-dimensional random sampling designs commonly exist for contamination sampling over space [12]. According to Jessen [13] which proposed high accuracy can be improved by better sampling and procedures, so samples collected with low bias and high precision from the centre of contamination zone.

Soil Sampling and Site Assessment

Determining the extent and spatial distribution of NAPL ensures that the chemical flood targets the proper subsurface volume. There are many techniques that investigators have used to map NAPL distribution. Perhaps the most precise and appropriate parameter to measure characterizing an NAPL source zone is saturation.

Saturation is defined as the fraction of the soil pore volume that is filled with NAPL. Depending on the degree of saturation, NAPL will exist as either free-phase NAPL or as residual NAPL. Free-phase or mobile NAPL exists when the saturation is high enough to form pore-to-pore connections over a large area, producing a continuous fluid capable of flowing under an imposed gradient or its own gravitational potential. Residual or entrapped NAPL exists when the soil pores have been drained of mobile NAPL, leaving behind some amount of liquid trapped by capillary forces or the surface tension that holds a liquid to a solid surface. NAPL at residual saturation is discontinuous and immobile, unable to flow under normally imposed hydraulic gradients.

A sampling technique was developed for allowing the decision makers to determine the minimum number of field realizations necessary to achieve a reliable remediation design [14]. This section provides collecting and analyzing soil samples from the NAPL zone. Soil samples can be used to provide an estimate of how NAPL is distributed in the source zone by providing contaminant concentration data that can be converted to saturation estimates. Soil samples also provide an indication of the vertical heterogeneity in the zone of interest. Laboratory analyses will yield a measurement of the total concentration in the soil samples. Soil samples are also collected to measure grain-size distribution, and the fraction of organic contaminant of the geologic media. Grain-size analyses are used to define heterogeneity and point permeabilities in the subsurface and should be collected at a frequency sufficient to define the major hydrostratigraphic units in the NAPL zone [15].

Representative soil samples were collected from different locations: (A) Helwan cement Co. El-Minia factory site, (B) PEPSI Cola El-Minia factory site, (C) Egyptian Co. for productive Electricity site, and (D) Middle Egypt Mills Co. site, as shown in the map Figure 2, and the lithological succession shown in Figure 3. In the selected sites, the contaminant (waste lubricant oils) distribution on the surface was observed to be inconsistent. Visual observation and random estimates of the oil content indicated variety of infiltration rates up to a depth of 40 cm. Accordingly, three samples, weighing 1 kg each, were collected from different depths going upto a maximum of 40 cm from the top layer beneath the ground surface [16]. Samples were stored in plastic buckets and homogenised in a mixer and it was air-dried for 3 days before selecting a sample for analysis and cleaned for sieving [17]. Size classification was achieved by sieving into gravel, medium, coarse, fine, and very fine sand also silt + clay fractions.

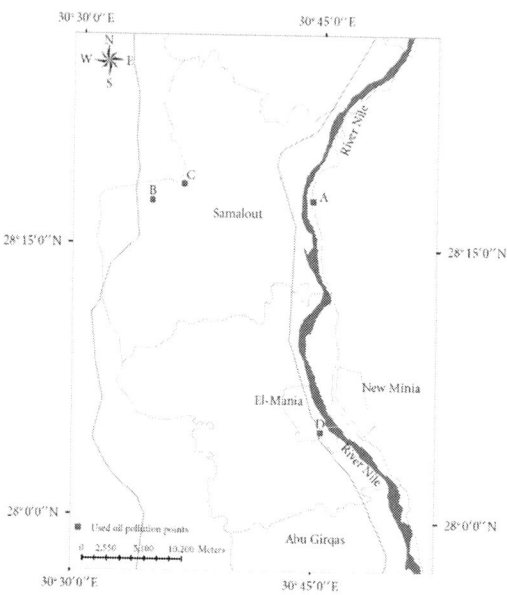

Figure 2: Location map of some oil polluted sits in Minia governorate.

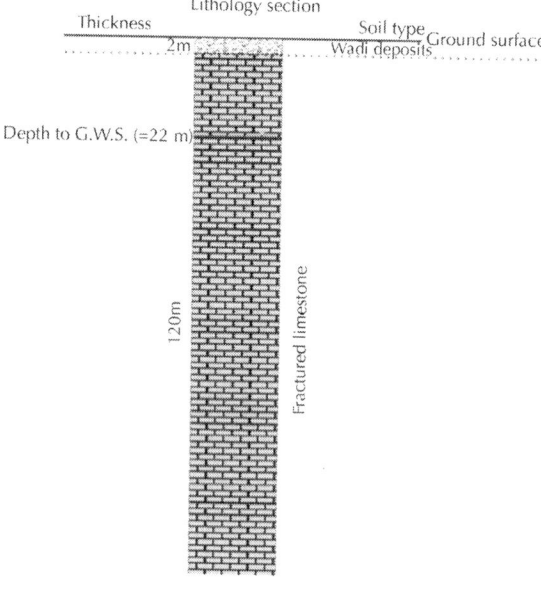

(a)

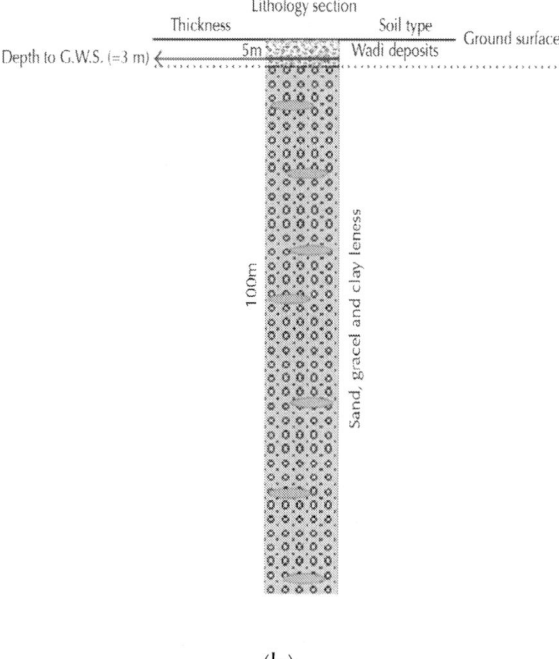

(b)

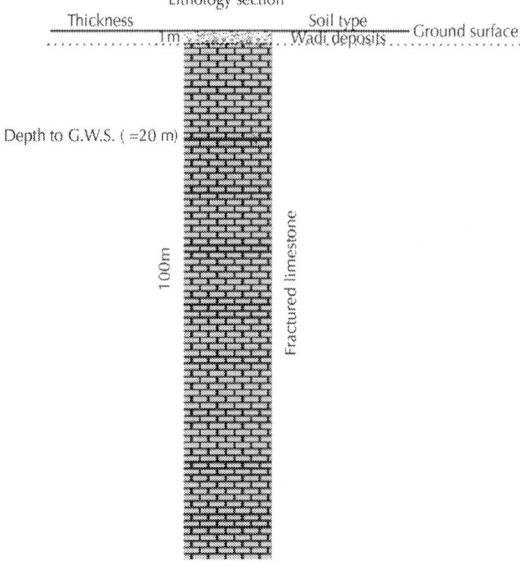

(c)

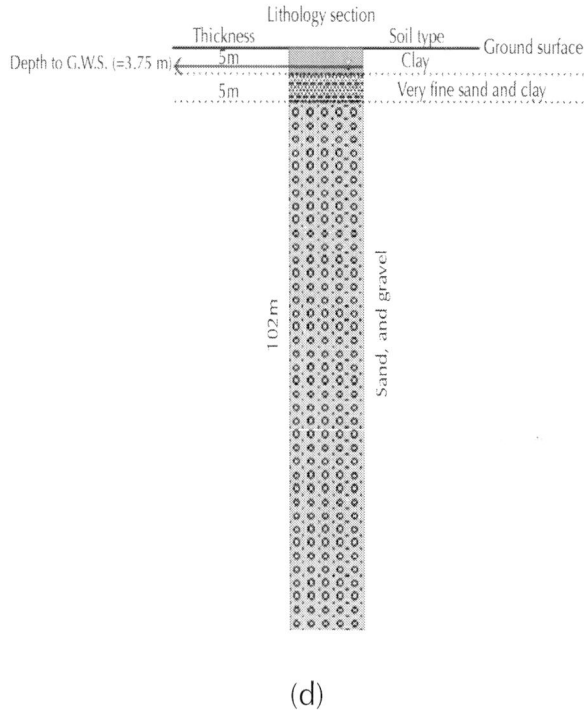

(d)

Figure 3: Lithology section of the investigated area [18].

MATERIALS AND METHODS

Materials

- Commercial nonionic surfactant nonyl phenol ethoxylate ($NPEO_{9.3}$) was used without treatment.
- The n-hexane used for extraction step was obtained as analytical grade solvent of used oil from the soil.
- Polluted soils from different sites in El-Minia governorate were collected to study the actuality of contamination problem and efficacy of treatment trains. Physical properties of the used lubricating oil are given in Table 1.

Table 1: Physical properties of the used lubricating oil

Specific gravity at 20°C	0.875
Flash point, close cup Pensesky Martin (°C)	140
Water and sediment (vol%)	1.28
Water content, the Dean and Stark method (vol%)	0.79
Viscosity at 37.8°C, cst	209.235
Ash content (wt%)	0.714
Asphaltene content (wt%)	4.995

EXPERIMENTAL SETUP

The actual contamination field conditions is simulate and designed in two experimental set-up models and as shown in Figures 4 and 5. The downflow mode was applied in this experiment.

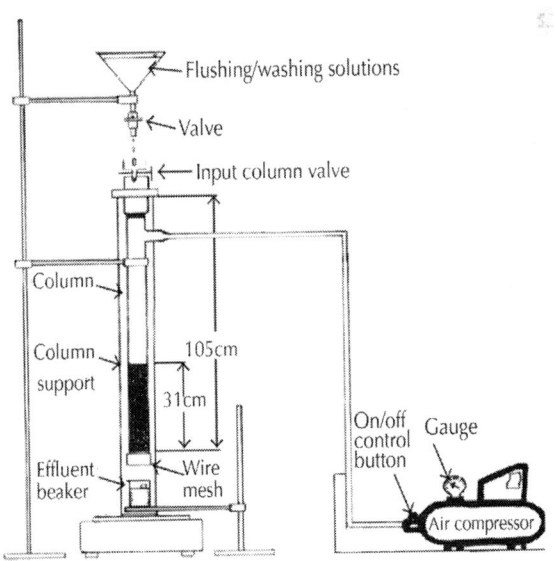

(a)

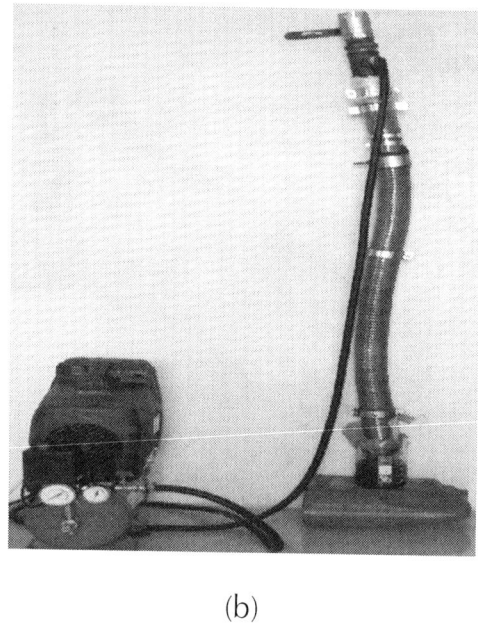

(b)

Figure 4: The laboratory experimental model simulates injection well.

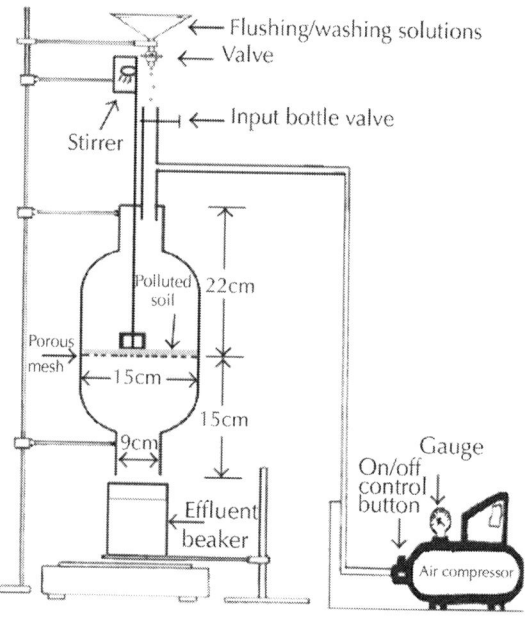

(a)

(b)

(c)

Figure 5: The laboratory experimental model simulates treatment tank.

System Operation

First Laboratory Model

The first lab model simulate injection well was consisted of a cylindrical column having dimensions (105 cm × 5 cm × 4.5 cm) caped with a stainless steel valve top. The outlet end of the column was fitted with a fine wire mesh screens (50 μm diameter) to prevent soil from washed out, then 31 cm of contaminated soil (1 kg) was packed in the column filled with surfactant solution to avoid entrapped air. After that the air was injected using air compressor to provide air and to pump surfactant solution and water across contaminated soil. The air was inlet from the side of the column to create turbulent air current, to provide better distribution of the injected air with similar and strong focusing at all points of soil surface. Air-flushing systems operated by pulsed mode (i.e., turning the system on and off at specified intervals), the pulsed air (cyclical) injection intermittently is thought to achieve more extensive, better distribution, and mixing of the air in the contaminated saturated zone, thereby allowing greater contact with the dissolved phase contaminants [19]. A fixed 300 mL of nonyl phenol ethoxyl ($NPEO_{9.3}$) surfactant solution of 3, 5 and 7% concentrations were injected individually at constant (2 bars) air pressure into different polluted field samples having concentrations of 2.55%, 12.7%, 5.8%, and 4.37% for loamy sand (site A), sand (site B), loam (site C), and clay loam (site D), respectively.

Second Lab Model

The second lab model simulates treatment tank were designed to remediate field samples of sites C and D because they contain high percent of silt and clay respectively. The treatment tank consisted of bottle having dimension (37 cm × 15.5 cm × 15 cm). The pores membrane was located below the half of bottle and covered with a fine nylon mesh screens (60 μm diameter) to prevent emulsion formation due to presence of silt and clay having very small grain size range and to facilitate of their washing out during the experimental work. Then 31 cm of contaminated soil (1 kg) was packed in the bottle. After that the air was injected using air compressor with a pressure not reach to 2 bar. Each samples were stirred with water after surfactant was washing out and then subjected to air pressure at 1.5 bar for two minutes, this step was repeated twice using 4 liters of water to wash out silt or clay that restrict air injection, emulsion extraction, and treatment process.

Treatment of Contaminated Soils

For all experiments the treatment of contaminated soils was carried out in two steps: flushing with surfactant solution then washing by many polishing cycles of water. The effluent samples were collected in a beaker and kept at room temperature for analysis, also soil was taken after each run to evaluate the percentage of waste oil remaining in the soil after each run.

Determining Oil Removal

n-hexane (C_6H_{14}) was used to extract the waste oil as herein after by Khalladi et al. [20] and Marek et al. [21].

Extraction and Analysis

Samples under distinct extractions: oil in soil phases and oil in water phases.

Oil in Soil

2 g of rinsed soil were mixed with 10 mL of n-hexane in a glass test tube and shaken laterally for 5 min then left at rest for separation, then the n-hexane/oil extract was removed and the mixture was poured in a beaker. Each rinsed soil was washing four times to ensure

completely clean from residual oil, this process was repeated till the extract gave the same absorbance reading as a pure n-hexane (zero absorbance) by using ultra violet spectroscopy.

Oil in Water (Emulsion)

The remaining pollutant concentration was determined after stopping the water washing cycles. 5 mL of liquid effluent (oil-in-water emulsion) were collected and mixed with 25 mL of n-hexane in a separating funnel; stirred for 2 min then left at rest for separation; the upper organic layer was separated by using separating funnel.

RESULTS AND DISCUSSION

We need to define the vertical and lateral extent of contamination in soil to determine if active remediation is necessary (i.e., if in situ action will take place, and if so, to what degree). From there, the vertical and lateral extent of contamination must be delineated to a level that is at or below the subsurface soil risk-based screening level.

Measuring Pollutant Concentrations

For many industrial sites, pollutant concentrations are highly variable, especially if the pollutants are immobile and there are multiple phases of polluting activity. It is known that the procedure selected to inhibit spills on land will greatly vary with the amount and type of oil spilled also with the type of soil and the terrain. Less viscous oil and more porous soil will allow greater and more rapid penetration and lateral migration in the soil.

To verify that the polluted soil is not contaminated above acceptable limits, confirmatory samples (discreet) will be taking from different sites A, B, C, and D. The soil samples are selected according to [22] who state that it is up to you to determine how many soil samples are needed to be representative of the conditions remaining in the site. The contaminated soils were selected by choosing four different location sites. Each site samples were taken separately from different depths (0 up to 40 cm). The proportion percentage of oil in each depth was then determined and added together and their average were taken and consider as representative sample for each site and defined by average waste oil concentration, and the samples analyzed based on our knowledge of waste lubricant oil and initial sample results are given in Table 2.

Table 2: Waste lubricant oil concentrations in soil samples from A, B, C, and D sites

Samples locations	Samples Classification	Waste oil concentrations from 0–10 cm (g/kg)	Waste oil concentrations from 10–20 cm (g/kg)	Waste oil concentrations from 20–40 cm (g/kg)	Average waste oil concentrations (g/kg)
Site A	Loamy sand	35.70	22.67	18.14	25.50
Site B	Sand	130	125.58	125.58	127.05
Site C	Loam	60.32	58.64	55.29	58.08
Site D	Clay loam	52.96	43.51	31.73	42.73

The results in Table 2 indicate that the average residual oil concentrations in contaminated site A, C, and D caused by movement of oil mobile phase (free phase) and thus their concentrations in the soil are not homogeneous, and recorded the following values 2.5%, 5.8%, and 4.2%, respectively. Also the pollutant concentrations at sampling points, or indeed from within layers of similar appearance, show heterogeneous nature of contaminant distribution. The oil percentage in site B is 12.7% this means that such site existing in continuous region due to high pollutant concentration. From all results we can conclude that the processes of adsorption-desorption play important roles in controlling the migration rate as well as concentration distributions. These processes tend to retard the rate of contaminant migration and act as mechanisms to reduce concentrations.

Effect of Soil Heterogeneity versus Pollutant Extraction

First lab model (Figure 4) that simulates injection well was used for treatment of all representative polluted samples sites (A, B, C, and D). Results of treatment site A and site B are given in Tables 3 and 4, respectively. The data indicated that only those sites are completely fulfilled and take action towered treatment by injection well simulated lab model. This can be attributed to that the sample from site A is loamy sand, while sample from site B are sand, thereby these types of soil are suitable to treated in site (in situ) during injection and subsequently extraction wells. Our conclusion is identical with Okx and Stein [23] who stated that it's possible to treatment the polluted soil from gravel to fine sand by in situ methods. On the other hand they confirmed the impossibility of in situ treatment for soil samples from sites C and D because their analysis indicated that they classified to Loam and clay loam types. To solve this problem the author suggest to go behind Boelsma et al. [24] who used Figure 6 to discussed how far the pressurized liquid extraction can be applied with different soil texture and they mentioned that its suitable to use pressurized liquid extraction technology for sandy clay, sandy clay loam, sandy loam, loamy sand, and sand, and they represented by shaded areas at which pressurized liquid extraction is typically effective. Therefore, the pressurized liquids technique that was used in current study is considered. Grain size analyses of the samples in current study have been determined using the standard sieving (>2–<0.004 mm), the results of grain size distribution of the current samples (A–D) are represented in histograms Figure 7. The grain size values reveal that contaminated soil samples that treatment by pressurized liquid extraction (i.e., use of injection solutions) are represent different soil type. These soils are loamy sand, sand, loam and clay loam, for the site A, B, C, and D, respectively. Soil heterogeneity plays a major role in controlling the concentration of contaminants extracted from the contaminated soil. It was also found that the efficiency of remediation depended on the type of the soil and it was much higher for the sand than for the clay soil. This difference can be explained by much looser structure of the soil particulates in the sandy soil and much higher stickiness and plasticity of clay [25]. Clay soil is plastic, consisting mainly of hydrous silicate of aluminum. At the microscopic level, clay is composed of fine particles (diameter <2 µm), adhering easily to one another [2]. The effect of pressurized liquid extraction on the removal of waste oil are calculated according to Couto et al. [26] and expressed as in (1),

$$\eta (\%) = \frac{m_o}{m_{oR}} \times 100. \tag{1}$$

where (η) is a remediation efficiency at any time, m_o is the total amount of oil removed by the remediation fluids in a given period of time, and m_{oR} is the original mass of oil in the soil.

Table 3: Results of soil treatment from site A

Flushing cycles	Pulsing time from zero-2-zero bar (minute)	(m_{oR}) original weight (g/kg)	(m_o) amount of oil removed (g/kg)	% of oil extracted
3% NPEO$_9$+ first air pulse	91	25.5	9.1	35.69
Water + second air pulse	80	25.5	8.39	32.90
Water + third air pulse	60	25.5	2.01	7.89
Water + fourth air pulse	51	25.5	1.55	6.08
Water + fifth air pulse	51	25.5	0.95	3.73

Water + sixth air pulse	51	25.5	0.91	3.57
Water + seventh air pulse	51	25.5	0.87	3.41
Total results of site A	435	25.5	23.78	93.26

Table 4: Results of soil treatment from site B

Flushing cycles	Pulsing time from zero-2-zero bar (minute)	(m_{oR}) original weight (g/kg)	(m_o) amount of oil removed (g/kg)	% of oil extracted
First group set				
3% NPEO$_9$ + first air pulse	25	127.05	10.72	8.44
Water + second air pulse	17	127.05	5.15	4.05
Water + third air pulse	14	127.05	4.33	3.41
Water + fourth air pulse	10	127.05	4.02	3.16
Water + fifth air pulse	9	127.05	3.28	2.58
Water + sixth air pulse	8	127.05	3.05	2.40
Water + seventh air pulse	8	127.05	2.98	2.35
Sum of first group set	91	127.05	33.53	26.39
Second group set				
3% NPEO$_9$ + second air pulse	21	127.05	12.66	9.96
Water + eighth air pulse	12	127.05	6.08	4.79
Water + ninth air pulse	12	127.05	5.54	4.36
Water + tenth air pulse	10	127.05	5.22	4.11
Water + eleventh air pulse	10	127.05	5.19	4.09
Water + twelfth air pulse	10	127.05	5.03	3.96

Sum of second group set	75	127.05	34.69	27.30
Third group set				
3% $NPEO_9$ + third air pulse	35	127.05	25.53	20.09
Water + thirteenth air pulse	3	127.05	3.12	2.46
Water + fourteenth air pulse	2	127.05	3.04	2.39
Water + fifteenth air pulse	2	127.05	2.88	2.27
Water + sixteenth air pulse	1.45	127.05	2.61	2.05
Water + seventeenth air pulse	1.45	127.05	2.99	2.35
Water + eighteenth air pulse	1.45	127.05	2.5	1.97
Water + nineteenth air pulse	1.45	127.05	1.44	1.13
Water + twentieth air pulse	1.45	127.05	1.39	1.09
Water + twenty-one air pulse	1.45	127.05	1.3	1.023
Water + twenty-two air pulse	1.45	127.05	1.26	0.99
Water + twenty-three air pulse	1.45	127.05	1.13	0.89
Water + twenty-four air pulse	1.45	127.05	0.98	0.77
Water + twenty-five air pulse	1.45	127.05	0.59	0.46
Sum of third group set	56.5	127.05	50.76	39.95
Total results of site (B)	222.5	127.05	118.98	93.65

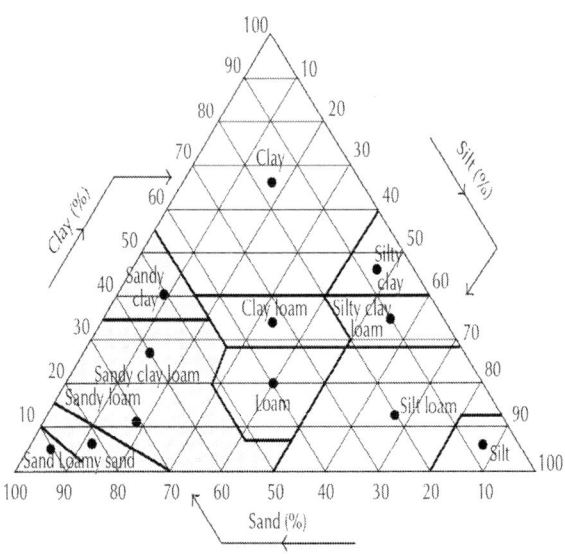

Figure 6: Triangular diagram illustrate the soil texture classification [37].

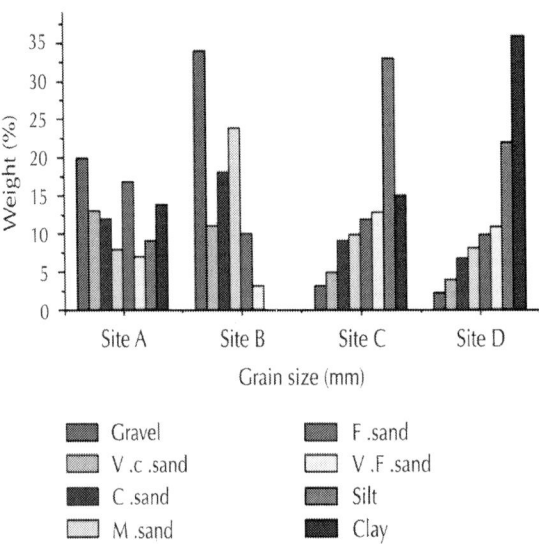

Figure 7: Histogram illustrates the grain size distribution of the studied polluted samples.

In Situ Methods for Treatment of Contaminated Soils

The representative sample from site A was found to be loamy sand soil texture and containing 2.55% (25.50 gram/kg) waste lubricant oil. The results in Table 3 reveal that, the percentage of removed oil is significantly increased with decreasing air pulsing time. That is, the first air pulse removed out 35.86% of waste lubricant oil recorded after 91 minutes and that time is decreased to 51 minutes beginning from the fourth washing cycle and fixed at that value till the seventh run is finished.

Also, it is found from the first surfactant solution run to the sixth water washing run the time of pulsed emerge nearly decrease to half. Such drastic increase in pressure time drop is attributed to decrease in aqueous phase, which results from cleaning or opening of air channels [27, 28].

The waste lubricant oil concentrations in site B is 12.70% (127.05 gram/kg) and is classified as soil sand texture as given in Table 2. The oil removing from site B is carried out using injection well-simulated lab model and is divided to main three air-flushing cycles with 3% surfactant solution followed by different water washing cycles runs, so the total washing water cycles runs reached to twenty-five. The results in Table 4 reveal that at the first group 8.43% of oil was removed after surfactant flushing and that values was decreased to 3.08% after the sixth water run was finished, that is, till the emulsion created during surfactant flushing have been completely push out. On the other hand the total waste oil removal from the first flushing group set is 29.82% recorded after surfactant flushing followed by six water washing cycle and at total time equal 91 minute.

This result is not satisfied due to the degree of saturation and age of waste oil in this sample not solubilized yet. This deduction matched with Zhou and Rhue [29] they confirmed that, the solubility of hydrocarbons depends on the type and quantity of surfactant, and the age of contamination. Beck et al. [30] emphasized the biphasic nature of contaminant release from the solid phase; where compounds are not degraded or lost fairly quickly from soil, their chemical and biological availability decreases rapidly over periods of minutes to hours, then slowly over periods of weeks to months through sorption and diffusion processes referred to as "ageing" [31].

In the second flushing set group is carried out on the remaining soil using 300 mL of 3% surfactant solution followed by five water washing cycles and the results are given in Table 4. The results reveal that, the percentage of removed oil was 9.96% from residual oil after first surfactant flushing run and decreased to 4.79% for next run (water washing cycle). The later water washing cycles remove oil out from 4.36% to 3.96%. On the other hand 27.30% of residual waste oil after the first treatment run was extracted with total time equal 75 minute. It is clear that the percentage of oil recovery proceed in reverse order with adsorption process (adhesion of oil on the surface of the soil). Sorption has a rapid phase within 48 h and a slower desorption phase that can take weeks even years [32]. In response, the effort is to quickly recover or cleanup spilled waste lubricant oil both from the ground surface and from subsoil. Reducing of waste lubricant oil concentrations in the soil below those corresponding retention capacities (saturation values) by fast recovery will stop spreading of waste lubricant oil in soil dramatically.

For the third flushing cycle, as revealed in Table 4 the results indicated that 20.09% was pushed out during this flushing run with air pulse time recorded 35 minutes for emerge. This means that solubility originated coherently with the third flushing solution. Also the time dropped from 35 minute where stable at 1.45 minute from eighteenth to twenty-seven air pulse. Such drastic increase in pressure time drop is attributed to decrease in aqueous phase, which results from cleaning or opening of air channels as mentioned by Clayton [27] andChao et al.[28]. Finally, the total waste lubricant oil recovery of site B is 93.65%.

Effect of Treatment Time and Soil Heterogeneity-related Issues

The results in Table 5 reveal that the total flushing durations are 91 and 81 min, total washing durations are 344 and 141.5 mins for sites A and B, respectively. The high flushing and washing duration time can be attributed to the high percentage of mud in site A over site B, that is, 19% and zero, respectively. This means that the performance of surfactant flushing/water washing can be adversely affected by geologic heterogeneity, because soil heterogeneities can cause poor-pressurized liquids sweep of the area targeted for remediation, although heterogeneity may reduce effectiveness. With increasing silt and clay content in polluted soil, the corresponding treatment duration time has been directly increased as showing in Figures 8 and 9. Thus, pressurized liquid technologies work in heterogeneous media, but cleanup times will be longer and more difficult to estimate than for similar systems in more homogeneous media.

Table 5: Comparison between treatment conditions related to site A and site B samples

Sample	No. of NPEO $_9$ flushing cycles	No. of water washing cycles	Total flushing duration (minute)	Total washing duration (minute)	Mud (silt + clay) (%)
Site A	1	6	91	344	19
Site B	3	24	81	141.5	0

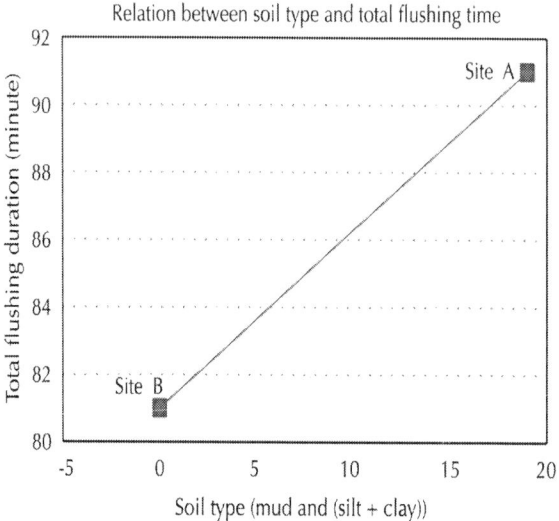

Figure 8: The relationship between mud % and surfactant flushing time (minute).

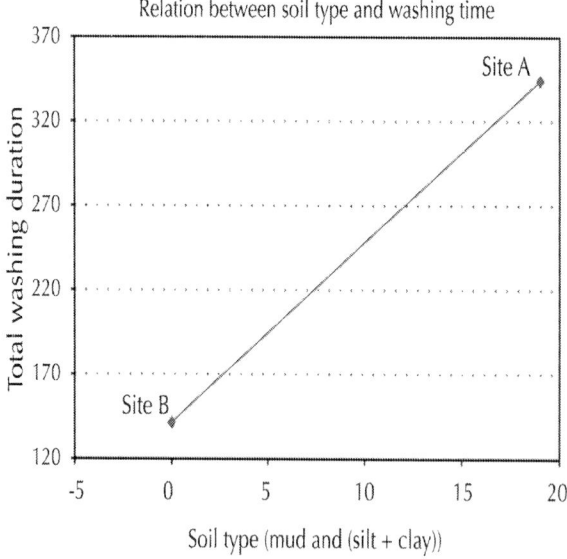

Figure 9: The relationship between mud % and water washing times (minute).

In general, sites having high clay or silt content in soils are not typical in situ candidates for this technology [33]. In an environment with low permeability layers or units interspersed with higher permeability zones, fluids preferentially flow through higher permeability zones, thereby reducing the performance of the chemical injected in the lower permeability zones. In general, high permeability soil is favored for pressurized liquids since numerous pore volumes can be passed through the contaminated area. Layered systems are difficult to remediate due to limited contact in less permeable areas. Also, fractured rock is extremely challenging due to the flow complexity in the fractured media. However, these challenges are relevant for all remediation technologies. Comparisons under the same conditions in these difficult media should be performed determine if surfactant flushing is able to remove contaminants faster than other technologies [34].

Ex Situ Methods for Treatment of Fine-grained Contaminated Soils

As mentioned before and illustrated in Figure 6, both of sites C and D characterized by loam and clay loam could not able to applying the injection well simulation lab model (in situ treatment), therefore another lab model has been used and known by simulates treatment tank (ex situ treatment) and its outline is given before in Figure 5. The (ex situ treatment), that is, sediment relocation has been used internationally as an operational response to treat oil spills throughout the world [35]. Ex situ involves the excavation of contaminated soil for treatment in aboveground. The soil may be transported to special facilities where remediation may be carried out in special reactors or vessels, which are specially designed for this purpose (ex situ or in tank method) [36]. An example of this process is the washing of heavily polluted soils in special tanks. The polluted soil may also be transported and spread on a surface prepared to prevent the spread of contamination in lateral and vertical directions. Beds arranged in this way form the so-called prepared beds, upon which the remediation process will take place.

This method is especially suitable for soils contaminated with oil products.

Ex situ treatment have been applied for treatment polluted soil tacking from site C that containing original weight 58.08 g/kg waste lubricant oil and characterized by loam soil texture. The results in Table 6 reveal that the percentage of removed oil is 84.99% after first flushing cycle using 3% nonionic surfactant solution. This high percentage of oil removal has been realized under low pressure equal 1.5 bar and at very short time duration about 2 minutes. By the same manner at the same time and pressure the second air injection (water washing cycle) was applied to push out 13.46% from residual oil. So the total percentage of extracted waste lubricant oil rises to 98.45%. Both flushing and washing cycles follow on stir periods to wash out very fine grains which subjected to measure the grain size diameter via petrographic microscope and the photomicrograph are given in Figure 10. Such figure represent photomicrograph of higher clay content and polluted oil patches in loam soil of site C sample are ranged from 0.063 mm to less than 5 µm.

The same treatment manner has been applied to treatment of site D sample that containing 42.73 g/kg original weight wastes lubricant oil and characterized as clay loam soil texture. The results in Table 7 reveal that the percentage of removed oil is 44.35% after first flushing cycle using 3% nonionic surfactant solution under pressure of 1.5 bar and duration time 2 minutes. The percentage of oil removal in the second air injection (water washing cycle) is 27.57% from residual oil. So the total percentage of extracted waste lubricant oil rises to 71.92%. Both flushing and washing cycles follow on stir periods to wash out very fine grains and after that the sample was subjected to measure the grain size diameter via petrographic microscope and the photomicrograph are given in Figure 11. Such figure represent the that the photomicrograph of higher clay content and oil pollution patches in clay loam soil of site (D) sample are ranged from 0.063 mm to less than 10 µm.

Table 6: Results of soil treatment form site (C)

Flushing cycles	Pulsing time from zero-2-zero bar (minute)	(m_{oR}) original weight (g/kg)	(m_o) amount of oil removed (g/kg)	% of oil extracted
3% NPEO$_9$ + first air pulse	2	58.08	49.36	84.99
Water + second air pulse	2	58.08	7.82	13.46
Total results of site (C)	4	58.08	57.18	98.45

Table 7: Results of soil treatment from site D

Flushing cycles	Pulsing time from zero-2-zero bar (minute)	(m_{oR}) original weight (g/kg)	(m_o) amount of oil removed (g/kg)	% of oil extracted
3% NPEO$_9$ + first air pulse	2	42.73	18.95	44.35
Water + second air pulse	2	42.73	11.78	27.57
Total results of site (D)	4	42.73	30.73	71.92

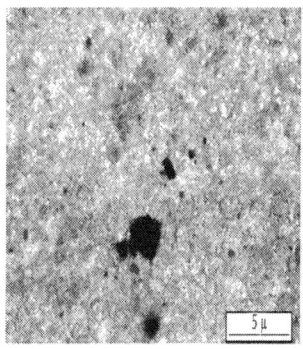

Figure 10: Photomicrograph show the higher clay content and oil pollution patches in loam soil of site C sample.

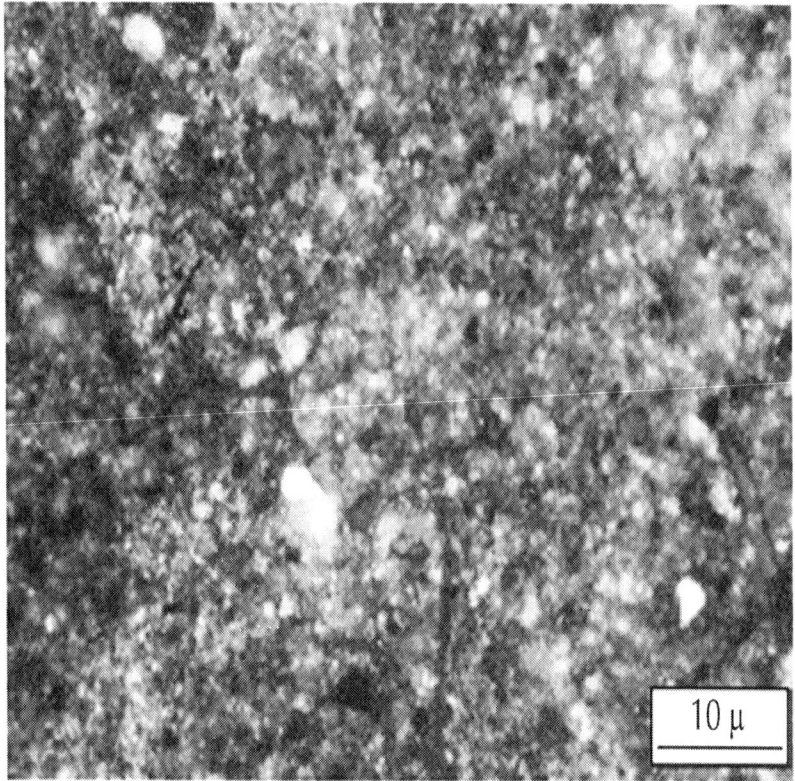

Figure 11: Photomicrograph shows the higher clay content and a lot of oil patches in clay loam soil of site D sample.

The efficiency of detoxification (i.e., waste lubricant oil removal) rate and extent for a waste is a function of soil type arises in Tables 6 and 7. The data reveals that the loam soil give satisfied results, where clay loam gives worse results, this because of its higher stickiness and plasticity. Also clay loam soil (site D) contains swelling clay that reduces permeability and emulsion formation, as mentioned by Kujawski et al. [2]. The swelling clay results from absorption of aqueous solution (water) which causes loss of surfactant due to soil sorption, sorbed surfactant molecules of the solids also increase contaminant sorption. This phenomenon was observed for 2-methynaphthalene [38], PCP [39], and other compounds. Our results are in agreement with Kibbey and Hayes [40] they mentioned that loss of aqueous surfactant due to soil sorption may significantly increase the surfactant doses required to enhance site remediation.

Generally, in the ex situ process the removal efficiency can be better controlled and the cleanup period is relatively short. This accomplished during soil treatment process of site C and site D samples, which carry out through separates fine soil (clay and silt) from coarse soil (sand and gravel). Since hydrocarbon contaminants tend to bind and sorb to smaller soil particles (primarily clay and silt), separating the smaller soil particles from the larger ones reduces the volume of contaminated soil. Soil washing depend on clay and silt separation processalso recommended by Riser-Roberts [41]. Also soil treatment in aboveground reactors allows a greater process control that is generally impossible with in situ techniques. Mass transfer of organic compounds (desorption from solids) is greatly increased in ex situ remediation process because its more applicable to fine-grained, low permeability soils which are not amenable to in situ techniques. In general, ex situ remediation is favored over in situ techniques for heavily contaminated soil and relatively localized and shallow contamination [42]. Ex situ soil washing is commonly used for treating contaminated soils by separating the most contaminated fraction of the soil for disposal. Surfactant enhanced ex situ soil washing can offer the convenience, efficiency and economy desirable for innovative and alternative soil washing technologies. However, surfactant selection guide-lines are needed to be evaluated in ex situ soil washing is important for the soil remediation industry [43].

On the other hand, RAAG [44], Chu and Chan [45] reported that the smaller volume of soil, that contains the majority of clay and silt particles, can be further treated by other methods (such as bioremediation) or disposed in accordance with environmental regulations, whereas to clean the larger volume of soil is considered to be nontoxic and can be used as backfill.

CONCLUSIONS

Soil remediation by surfactant solutions flushing is a common practice successfully done in a pilot-scale and field-scale studies, where the apparent contaminant solubility in surfactant solutions can be hundreds to thousands times higher than its water solubility if a traditional pump and treat method is applied. This example of the use of a treatment train for creosote-contaminated soil, and if applied either in suit or ex situwere involves: (1) free product removal as it possible using a pumping system, (2) flushing with surfactants and washing by water using air-sparging technology as distribution system and to insure kinetic energy for emulsion formation and moving, and (3) biodegradation of the residual contamination, by air sparging or air stripping.

Studied Soil Reach

The research work carried out throughout this study at Water laboratory Geology Department, El-Minia University, Faculty of Science, El-Minia, Egypt, and divided into two main parts: the first part is a laboratory study started since April 2009 ending July 2009, in this respect a simulated lab model is designed and packed with artificial polluted soils submerged with nonionic surfactant in presence of air sparging as mentioned in experimental part, where the second part is started in August 2009 ending in January 2010 by applying the results of the first part on some rails-polluted areas located in El-Minia, the results are not mentioned in this article, as well as they are not published or sent for publication till now.

ACKNOWLEDGMENTS

This work was supported by a grant from Egyptian Petroleum Research Institute, El-Minia University, Faculty of Science, and Geology Department and Engineering Petrotread Co.

REFERENCES

1. J. H. Harwell, D. A. Sabatini, and R. C. Knox, "Surfactants for ground water remediation," Colloids and Surfaces A, vol. 151, no. 1-2, pp. 255–268, 1999.
2. W. Kujawski, I. Koter, and S. Koter, "Membrane-assisted removal of hydrocarbons from contaminated soils-laboratory test results," Desalination, vol. 241, no. 1–3, pp. 218–226, 2009.
3. B. K. Gogoi, N. N. Dutta, P. Goswami, and T. R. K. Mohan, "A case study of bioremediation of petroleum-hydrocarbon contaminated soil at a crude oil spill site," Advances in Environmental Research, vol. 7, no. 4, pp. 767–782, 2003.
4. C. W. Fetter, Contaminant Hydrogeology, Prentice-Hall, Upper Saddle River, NJ, USA, 2nd edition, 1999.
5. J. F. Pankow and J. A. Cherry, Dense Chlorinated Solvents and other DNAPLs in Ground-Water, Waterloo Press, Portland, Ore, USA, 1996.
6. R. A. Mackay, in Nonionic Surfactants: Physical Chemistry, M. J. Schick, Ed., vol. 23 of Surfactant Science Series, pp. 297–367, Marcel Dekker, New York, NY, USA, 1985.
7. R. N. Yong, A. M. O. Mohamed, and B. P. Warkentin, Principles of Contaminant Transport in Soils, Elsevier, Amsterdam, The Netherlands, 1992.
8. A. Wild, Soils and the Environment: An Introduction, Cambridge University Press, 1993.
9. J. W. Mercer and R. M. Cohen, "A review of immiscible fluids in the subsurface: properties, models, characterization and remediation," Journal of Contaminant Hydrology, vol. 6, no. 2, pp. 107–163, 1990.
10. Y. J. Tsai, F. C. Chou, and S. J. Cheng, "Using tracer technique to study the flow behavior of surfactant foam," Journal of Hazardous Materials, vol. 166, no. 2-3, pp. 1232–1237, 2009.

11. D. Feng and C. Aldrich, "Sonochemical treatment of simulated soil contaminated with diesel," Advances in Environmental Research, vol. 4, no. 2, pp. 103–112, 2000.
12. R. O. Gilbert, Statistical Methods for Environmental Pollution Monitoring, Van Nostrand Reinhold, New York, NY, USA, 1987.
13. R. J. Jessen, Statistical Survey Techniques, John Wiley & Sons, New York, NY, USA, 1978.
14. X. Zhang, G. H. Huang, Q. Lin, and H. Yu, "Petroleum-contaminated groundwater remediation systems design: a data envelopment analysis based approach," Expert Systems with Applications, vol. 36, no. 3, pp. 5666–5672, 2009.
15. M. Vukovic and A. Soro, Determination of Hydraulic Conductivity of Porous Media from Grain-Size Composition, Water Resources, Littleton, Colo, USA, 1992.
16. S. K. Chaerun, K. Tazaki, R. Asada, and K. Kogure, "Bioremediation of coastal areas 5 years after the Nakhodka oil spill in the Sea of Japan: isolation and characterization of hydrocarbon-degrading bacteria," Environment International, vol. 30, no. 7, pp. 911–922, 2004.
17. C. K. Ahn, Y. M. Kim, S. H. Woo, and J. M. Park, "Soil washing using various nonionic surfactants and their recovery by selective adsorption with activated carbon," Journal of Hazardous Materials, vol. 154, no. 1–3, pp. 153–160, 2008.
18. RIGW, Feasibility of Vertical Drainage in the Nile Valley, Minia pilot Area, Ministry of Irrigation, Cairo, Egypt, 1986.
19. X. Yang, D. Beckmann, S. Fiorenza, and C. Niedermeier, "Field study of pulsed air sparging for remediation of petroleum hydrocarbon contaminated soil and groundwater," Environmental Science and Technology, vol. 39, no. 18, pp. 7279–7286, 2005.
20. R. Khalladi, O. Benhabiles, F. Bentahar, and N. Moulai-Mostefa, "Surfactant remediation of diesel fuel polluted soil," Journal of Hazardous Materials, vol. 164, no. 2-3, pp. 1179–1184, 2009.

21. S. Marek, K. Martin, M. Martina, and R. Robert, "Soil flushing by surfactant solution: pilot-scale demonstration of complete technology," Journal of Hazardous Materials, vol. 163, no. 1, pp. 410–417, 2009.
22. HMWMD, Management Standards for Used Oil Transporters Guidance Document, Hazardous Materials and Waste Management Division, Colorado Department of Public Health and Environment, 1st edition, 2005, http://www.cdphe.state.co.us/hm/oilgen.pdf.
23. J. P. Okx and A. Stein, "An expert support model for in situ soil remediation," Water, Air, and Soil Pollution, vol. 118, no. 3-4, pp. 357–375, 2000.
24. F. . Boelsma, E. C. L. Marnette, C. G. J. M. Pijls, C. C. D. F. van Ree, and K. Vreeken, In Situ Air Sparging, A Technical Guide, Geodelft Environmental, Delft, The Netherlands, 1999.
25. A. G. Link, "Textural classification of sediments," Soil Texture Classifications, Sedimentology, vol. 7, pp. 249–254, 1966.
26. H. J. B. Couto, G. Massarani, E. C. Biscaia, and G. L. Sant'Anna, "Remediation of sandy soils using surfactant solutions and foams," Journal of Hazardous Materials, vol. 164, no. 2-3, pp. 1325–1334, 2009.
27. W. S. Clayton, "A field and laboratory investigation of air fingering during air sparging," Ground Water Monitoring and Remediation, vol. 18, no. 3, pp. 134–145, 1998.
28. K. P. Chao, S. K. Ong, and A. Protopapas, "Water-to-air mass transfer of VOCs: laboratory-scale air sparging system," Journal of Environmental Engineering, vol. 124, no. 11, pp. 1054–1060, 1998.
29. M. Zhou and R. D. Rhue, "Screening commercial surfactants suitable for remediating DNAPL source zones by solubilization," Environmental Science and Technology, vol. 34, no. 10, pp. 1985–1990, 2000.
30. A. J. Beck, S. C. Wilson, R. E. Alcock, and K. C. Jones, "Kinetic constraints on the loss of organic chemicals from contaminated soils: implications for soil-quality limits," Critical Reviews in

Environmental Science and Technology, vol. 25, no. 1, pp. 1–43, 1995.
31. K. T. Semple, A. W. J. Morriss, and G. I. Paton, "Bioavailability of hydrophobic organic contaminants in soils: fundamental concepts and techniques for analysis," European Journal of Soil Science, vol. 54, no. 4, pp. 809–818, 2003.
32. D. M. LaGrega, P. L. Buckingham, and J. C. Evans, in Hazardous Waste ManagementThe Environmental Resources Management Group, P. H. King and R. Elianssen, Ed., Civil Engineering Series, Mc-Graw-Hill, Singapore, 1994.
33. B. C. Kirtland and C. M. Aelion, "Petroleum mass removal from low permeability sediment using air sparging/soil vapor extraction: impact of continuous or pulsed operation," Journal of Contaminant Hydrology, vol. 41, no. 3-4, pp. 367-383, 2000.
34. ITRC, Technical and regulatory guidance for surfactant/cosolvent flushing of DNAPL source zones. (Interstate Technology & Regulatory Council) DNAPL-3, 2003, http://www.itrcweb.org/Documents/DNAPLs-3.pdf.
35. M. O. Hayes and J. Michel, "Factors determining the long-term persistence of Exxon Valdez oil in gravel beaches," Marine Pollution Bulletin, vol. 38, no. 2, pp. 92–101, 1999.
36. G. A. Sergy, C. C. Guénette, E. H. Owens, R. C. Prince, and K. Lee, "In-situ treatment of oiled sediment shorelines," Spill Science and Technology Bulletin, vol. 8, no. 3, pp. 237–244, 2003.
37. D. C. DiGiulio, U.S. EPA Robert S. Kerr Environmental Research Laboratory, personal communication, Ada, Okla, USA, 1989.
38. Y. P. Chin, K. D. Kimble, and C. Robin Swank, "The sorption of 2-methylnaphthalene by Rossburg Soil in the absence and presence of a nonionic surfactant," Journal of Contaminant Hydrology, vol. 22, no. 1-2, pp. 83–94, 1996.
39. S. K. Park and A. R. Bielefeldt, "Aqueous chemistry and interactive effects on non-ionic surfactant and

pentachlorophenol sorption to soil," Water Research, vol. 37, no. 19, pp. 4663–4672, 2003.

40. T. C. G. Kibbey and K. F. Hayes, "A multicomponent analysis of the sorption of polydisperse ethoxylated nonionic surfactants to aquifer materials: equilibrium sorption behavior," Environmental Science and Technology, vol. 31, no. 4, pp. 1171–1177, 1997.

41. E. Riser-Roberts, Remediation of Petroleum Contaminated Soil: Biological, Physical, and Chemical Processes, Lewis, Boca Raton, Fla, USA, 1998.

42. W. Admassu and R. A. Korus, "Engineering of bioremediation processes: needs and limitations," in Bioremediation: Principles and Applications, R. L. Crawford and D. L. Crawford, Eds., pp. 13–34, Cambridge University Press, Cambridge, UK, 1996.

43. S. Deshpande, B. J. Shiau, D. Wade, D. A. Sabatini, and J. H. Harwell, "Surfactant selection for enhancing ex situ soil washing," Water Research, vol. 33, no. 2, pp. 351–360, 1999.

44. RAAG, Evaluation of Risk Based Corrective Action Model, Remediation Alternative Assessment Group, Memorial University of NewfoundlandNF, St John's, Canada, 2000.

45. W. Chu and K. H. Chan, "The mechanism of the surfactant-aided soil washing system for hydrophobic and partial hydrophobic organics," Science of the Total Environment, vol. 307, no. 1–3, pp. 83–92, 2003.

Citations

CHAPTER 1

Alejandro Medel, Erika Méndez, José L. Hernández-López, et al., "Novel Electrochemical Treatment of Spent Caustic from the Hydrocarbon Industry Using Ti/BDD," International Journal of Photoenergy, Article ID 829136, in press.

CHAPTER 2

Beata Gabri , Aleksandra Sander, Marina Cvjetko Bubalo, and Dejan Macut, "Extraction of S- and N-Compounds from the Mixture of Hydrocarbons by Ionic Liquids as Selective Solvents," The Scientific

World Journal, vol. 2013, Article ID 512953, 11 pages, 2013. doi:10.1155/2013/512953.

CHAPTER 3

Zhongxiang Chen, Yibin Yan, Said S.E.H. Elnashaie, Catalyst deactivation and engineering control for steam reforming of higher hydrocarbons in a novel membrane reformer, Chemical Engineering Science, Volume 59, Issue 10, May 2004, Pages 1965-1978, ISSN 0009-2509, http://dx.doi.org/10.1016/j.ces.2004.01.046.

CHAPTER 4

Neha Patni, Pallav Shah, Shruti Agarwal, and Piyush Singhal, "Alternate Strategies for Conversion of Waste Plastic to Fuels," ISRN Renewable Energy, vol. 2013, Article ID 902053, 7 pages, 2013. doi:10.1155/2013/902053.

CHAPTER 5

Aurora Hernández Enríquez, Michael Binns, Jin-Kuk Kim, Systematic retrofit design with Response Surface Method and process integration techniques: A case study for the retrofit of a hydrocarbon fractionation plant, Chemical Engineering Research and Design, Volume 92, Issue 11, November 2014, Pages 2052-2070, ISSN 0263-8762, http://dx.doi.org/10.1016/j.cherd.2014.02.030.

CHAPTER 6

Bijayani Biswal, Sachin Kumar, and R. K. Singh, "Production of Hydrocarbon Liquid by Thermal Pyrolysis of Paper Cup Waste," Journal of Waste Management, vol. 2013, Article ID 731858, 7 pages, 2013. doi:10.1155/2013/731858.

CHAPTER 7

Th. Abdel-Moghny, Ramadan S. A. Mohamed, E. El-Sayed, Shoukry Mohammed Aly, and Moustafa Gamal Snousy, "Effect of Soil Texture on Remediation of Hydrocarbons-Contaminated Soil at El-Minia District, Upper Egypt," ISRN Chemical Engineering, vol. 2012, Article ID 406598, 13 pages, 2012. doi:10.5402/2012/406598.

Index

A
Aquatic ecosystem 204

C
Central Composite Design (CCD) 164
Circulating fluidized bed membrane reformer (CFBMR) 81, 84
Cracking process 123, 133

D
Denitrification of gasoline 52
Dens nonaqueous phase liquids (DNAPLs) 206
Differential thermogravimetry (DTG) 191

F
Fourier transform infrared spectroscopy (FTIR) 188, 194
Fractional Factorial Design (FFD) 147

H
Hydrodesulphurization (HDS) 52

I
Ionic liquid 52, 53, 54, 56, 57, 59, 60, 61, 62, 64, 65, 66, 68, 69, 70, 71, 72, 73, 75, 76, 77, 78, 79

K
Key performance indices (KPI) 144

L
Linear least squares (LLS) 164
Low-density polyethylene (LDPE) 186

M
Methyl t-butyl ether (MTBE) 205

N
Nonaqueous phase liquids (NAPLs) 204, 205

P
Permeability layer 229
Plastic solid waste 118, 120, 133
Plastic waste 118, 120, 122, 124, 127, 132
Plug flow reactor (PFR) 91
Polyethylene terephthalate 120, 123

Pyrolysis 124, 125, 126, 134, 135

R
Response Surface Methodology (RSM) 138, 142, 172
Retrofit Design Approach (RDA) 138, 141
Root Mean Square Error (RSME) 147

S
Sequential modular mode (SMS) 151
Simulated annealing (SA) 170

T
Thermogravimetric analysis (TGA) 190
Thermoplastics 118
Tightly bind 207
Total ion chromatogram (TIC) 195

W
Waste management strategy 117